# TIMESCAPES OF MODERNITY

Time is the invisible 'other' that works outside and beyond the reach of our senses. Thinking of the environment as a timescape allows us to see the hazards of an industrial way of life. The invisible becomes tangible and we begin to recognise processes that work below the surface until they materialise as symptoms – sometime, somewhere.

*Timescapes of Modernity* focuses on time to facilitate a deeper understanding of the interactions between environmental, economic, political and socio-cultural concerns. Barbara Adam argues that environmental hazards are inescapably tied to the successes of the industrial way of life: global markets and economic growth, large-scale production of food, the speed of transport and communication, the 24-hour society, even democratic politics. Emphasising the complexity of time, this book brings to the forefront of socio-environmental concerns the rhythms, timings, changes, latencies and contingencies that permeate the story of industrial success and excess.

Introducing a unique 'timescape' perspective, Barbara Adam dislodges taken-for-granted assumptions about environmental change, enables a reformulation of environmental problems and their cures, and provides the potential for innovative new strategies to deal with some of the most severe environmental hazards of our time.

**Barbara Adam** is Reader in Social Theory at the School of Social and Administrative Studies, University of Wales, Cardiff.

# GLOBAL ENVIRONMENTAL CHANGE SERIES
Edited by Jim Skea, University of Sussex

The *Global Environmental Change Series*, published in association with the ESRC Global Environmental Change Programme, emphasises the way that human aspirations, choices and everyday behaviour influence changes in the global environment. In the aftermath of UNCED and Agenda 21, this series helps crystallise the contribution of social science thinking to global change and explores the impact of global changes on the development of social sciences.

Also available in the series:

# TIMESCAPES OF MODERNITY

## The environment and invisible hazards

*Barbara Adam*

LONDON AND NEW YORK

First published 1998
by Routledge
2 Park Square, Milton Park, Abingdon, Oxon OX14 4RN

Simultaneously published in the USA and Canada
by Routledge
711 Third Avenue, New York, NY 10017

*Routledge is an imprint of the Taylor & Francis Group, an informa business*

Typeset in Garamond by RefineCatch Limited, Bungay, Suffolk

*British Library Cataloguing in Publication Data*
A catalogue record for this book is available from the British Library

*Library of Congress Cataloging in Publication Data*
Adam, Barbara
Timescapes of modernity: the environment & invisible hazards/Barbara Adam.
p.   cm. – (Global environmental change series)
Includes bibliographical references and index.
1. Human ecology. 2. Time. 3. Environmental economics.
4. Economic development – Environmental aspects. I. Title.
II. Series.
GF98.A33   1998
304.2–dc21      97–23332

ISBN 0–415–16274–2 (hbk)
ISBN 0–415–16275–0 (pbk)

# CONTENTS

# PLATES

# ACKNOWLEDGEMENTS

This research has been conducted during a two-year fellowship 1994–6 which was funded under the British Economic and Science Research Council's initiative on Global Environmental Change (L32027312593). The full-time funding enabled me to collaborate with colleagues in Germany and Austria and to participate in the development of a project – the Tutzing Time Ecology Project – that considers environmental issues from a temporal perspective. As a result of this collaboration I have drawn extensively on publications that have arisen from this work or are associated with it: *Politische Ökologie*, 'Vorsorgendes Wirtschaften. Frauen auf dem Weg zu einer Ökonomie der Nachhaltigkeit', Sonderheft 6 (1994); *Politische Ökologie*, 'Zeitfraß. Zur Ökologie der Zeit in Landwirtschaft und Ernährung', Sonderheft 8 (1995); M. Held and K. Geißler (eds) *Von Rhythmen und Eigenzeiten. Perspektiven einer Ökologie der Zeit* (1995). My research on the economics of the environmental timescape was significantly enhanced by the work of Martin Held and the recently deceased Bernd Biervert who over the last five years have organised seminars and published books on the ethical dimension of economics which included such topics as *Das Naturverständnis der Ökonomik* (The Economic Understanding of Nature), *Zeit in der Ökonomik* (Time in Economics) and *Die Dynamik des Geldes. Über den Zusammenhang von Geld, Wachstum und Natur* (The Dynamic of Money. On the Relation between Money, Growth and Nature). Among the numerous English language publications on the subject of environmental economics I found Michael Jacob's *The Green Economy* invaluable.

For Chapters 3 and 6 I draw on material researched during the fellowship that has been published in the following articles: 'Beyond the Present: Nature, Technology and the Democratic Ideal', *Time & Society*, 5,3:319–38 (1996c); 'Chernobyl: Implicate Order of Socio-environmental Chaos' in J. T. Fraser and M. Soulsby (eds) *Time and Chaos*, Madison, CT: International Universities Press (1997a); and 'Radiated Identities: In Pursuit of the Temporal Complexity of Conceptual Cultural Practices', in M. Featherstone and S. Lash (eds) *Spaces of Culture: City, Nation, World*, London: Sage (1997b). I would like to thank the publishers for giving their permission to use part of these texts.

ACKNOWLEDGEMENTS

Many people have supported me in this research: the people who talked to me about their images of nature, their approaches to 'the environment', their hopes and fears associated with pollution and environmental hazards; Lynda Whitehouse who has acted as my television spy and made sure that I was aware of all the programmes relevant to my research; family Hopkins who have collected newspapers for me; Miriam Adam who has once more scrutinised my references; Claude Siegenthaler who provided me with papers and references from the German speaking world of environmental and ecological economics; Maite Mathes who supplied me with equivalent information on geno- and biotechnology; and Martin Held with whom I had numerous discussions and who was tireless in his efforts to keep me informed across the breadth of time and ecology issues. To all these friends and colleagues I am deeply grateful. My sincere thanks also go to the readers of the first draft of this book for their insightful comments: Jan Adam, Stuart Allan, Martin Held, Beverly Holbrook, Caroline Joll, Judy Keenoy, Tom Keenoy, Jim Skea, Matthew Thomas and, most specifically, Andrew Coulson who has not only commented in detail on the penultimate version of entire text but also prepared the script for publication.

# INTRODUCTION

*Plate 1*  Ifor Tygwyn from *Wales the First Place* by Paul Wakefield. *Source*: Tony Stone Images.

## Troubled times

These are troubled times. As I write, the UK beef industry is collapsing, all fishing is suspended off the coast of West Wales and the tenth anniversary of Chernobyl has brought the invisible dangers of radiation back to the fore-front of public attention. Surrounded by so much disaster and tragedy it feels almost wrong and out of place to write about issues that I know to be important and deeply pertinent to this current round of environmental dis-asters. As friends and neighbours face existential crises, their livelihoods as farmers, fishermen, hoteliers, shopkeepers and artists seriously threatened by oil pollution and Bovine Spongiform Encephalopathy (BSE), I am forced to confront the relevance of my work not to the academic community of scholars but to the everyday lives of people caught up in the downward spiral of the industrial way of life. The grief and desperation in the area is palpable: a country and people stressed to their limits. West Wales, my adoptive 'home', has three principal sources of income: farming, tourism and fishing. All three have been deeply affected, threatened at the core by events that have arisen within one month of each other. Though concentrated in a par-ticular time–space, these events are symptoms of global economic and indus-trial processes which, in turn, are inseparably linked to specific conceptions and approaches to time and space, the subject matter of this book.

Where I was previously able to have long and intensive discussions about modern farming methods, about my farming friends' relationships with their animals, or innovations in 'the industry', today my work and my friends' existential fears stand in a problematic relation to each other. A shadow has crept into the long-standing relationship of friendship and trust. Aware of their despair and their hurt over the wide-spread 'farmer bashing', I find myself scrutinising and reflecting on everything I say, trying to see the issues simultaneously through their eyes. My desire to interview them feels inappropriate and opportunistic; to get their story and their views on both the causes and solutions of the beef crisis seems exploitative, insensitive to the gravity of the situation. In the midst of this tragedy, all I seem to be able to do is seek out and collate some of the less accessible information from the overload of stories provided by the media. Having followed events across the breadth of British newspapers and television I am able to pick out some of the thinly distributed morsels of conflicting factual information and pass them on like pieces of jigsaw for a still grossly incomplete puzzle.

My analysis of some of the temporal issues underpinning the crisis, how-ever, has to remain invisible, since this would add insult to injury in a situation where the combination of existential threat, scientific uncertainty and governmental bungling causes paralysis in the afflicted, preventing them from engaging in constructive thought and purposive action. How could I talk about the compression of time and the valorisation of speed in a social system that relates to time on an economic basis when increasing yields in ever shorter time constitute part of farmers' skills, when such temporal compression is a source of pride, and brings them respect in the farming community? Talk about what is wrong with industrial farming would not go down well at a time when farmers anticipate losing their herds, their basis of existence and their self-respect. How could I talk about the need for a renewed sensitivity to the natural rhythms of farm animals and their environment at a time when the whole cashflow system of beef and dairy farmers has collapsed, when they can no longer sell animals to pay for feed and fertilisers, when their patterns of exchange – their daily, weekly, monthly and seasonal rhythms of buying and selling – are no longer workable and when the debts mount proportional to the uncertainty not just of the immediate but also the long-term future? The shock of seeing their life-worlds implode and their livelihoods destroyed by a government so clearly incapable of handling the situation mixes helplessness with anger and hatred, disbelief with frustration and suspicion, soul searching with a desperate search for villains: the Europeans, the EU, the media, government, scientists, consumers, and agri-business which provides both the suspect feed and the machines for mechanical rendering. There is no need, it seems to me, to add 'armchair theorists' to that list. It would indeed be highly inappropriate for me to amplify their plight by showing links between BSE and approaches to time, between the symptoms of the industrial way of life and the persistent disregard of *Eigenzeiten*, that is, the system-specific times and seasonal rhythmicity of the live-stock and land in their care. And yet, to establish those links seems to me to be absolutely crucial.

## Knowledge for whom and what?

From the vantage point of hindsight these tragic events may well be considered as turning points, as opportunities for reflection and renewal. But somehow I do not see social theorists playing a central role in facilitating such changes. More than ever before, therefore, I am forced to question my work, query the purposes of my writing, confront my pride in scholarly activity. In the face of so much disaster and despair, I find it increasingly difficult to justify writing about conceptual matters for social scientists in the expectation that I affect their work and that this in turn will filter through to environmental policy and action. The need for theory and practice to be brought into the closest possible relation with each other seems ever more important: what I have to say should connect with and be relevant to the everyday lifeworld of work and social interaction, the concerns of policy makers and, most importantly, all the socio-environmental spheres of action and impact.

I have learnt much in this respect from my colleagues in Germany where I am part of a project that brings to the public's attention scientific work on the time aspects of environmental matters. In the form of publications and conferences, hosted by the education wing of the Protestant church with additional funds from the Schweisfurth Stiftung, Munich and, more recently, the Deutsche Bundesstiftung Umwelt, the 'Tutzing Time Ecology Project' brings together scientific work and lay knowledge (Held and Geißler, 1993; Held and Geißler, 1995; Schneider, 1995; Adam et al., 1997a). Here, academics are unashamedly programmatic and committed to present their work in a way that demonstrates the practical relevance of theory and scientific knowledge to the issues at hand. Moreover, they take account of the competencies of the lay public and are open to learning from the conference participants' specialist knowledge spheres. The subject matter of this project on time ecology may relate to agriculture, food, consumption, the treatment of water and soil, the changing pace of life, the economic perspective on the environment, the rhythmicities of nature and cities. Beyond this emphasis on praxis – the unification of theory and practice in the Marxian sense – the project puts into practice a theoretical commitment to embodied knowledge and aesthetic appreciation. Through a combination of science, art, music and poetry it presents as an integrated whole what institutional knowledge and professionalisation have set apart: academic knowledge, practical activity, embodied sense experience and aesthetic sensibilities. It thus fulfils its goal of truly embodied knowledge. Unlike the embodiment pursued in postmodern theory, however, this embodiment is not conceived as inscription but incorporation, not textual but practical, that is, tied to activity and praxis. (For an account of this project, see Adam et al., 1997b.)

I am further encouraged by the work of Michael Gibbons and his collaborators (Gibbons et al., 1994) on the new production of knowledge. They identify a new breed of knowledge production that is more contextual and use oriented, transdisciplinary and non-institutional, reflexive and socially accountable than was conceivable in the traditional academic mode. While Gibbons et al. give evidence of the production of such new social science, however, they show us little of the epistemological and ontological difficulties that are encountered in the shift from disembodied, decontextualised, institutional, objective science to openly acknowledged implication and explicit engagement. Yet we need to appreciate that the pursuit of this path creates new problems and revisits old ones that make the production of such social science far from easy. It necessitates change at the very centre of taken-for-granted scientific assumptions. Thus, to combine an appreciation of relativity as a fact of life (Adam, 1995a, Chapter 7) with a realist approach to nature, thorough-going reflectivity with a position of personal, embodied, contextual and critical engagement and, finally, disciplinary rigour with accessibility, demands dramatic revision of the social theory tradition. It necessitates a kind of social theory that, in his methodological writings, Max Weber (1949) – one of the founders of social science – designated impossible.

Over a much longer period, of course, feminist theorists have been leading the way towards such knowledge with their demand that theories have to become re-embodied and re-contextualised. Feminist scholars have consistently and persuasively argued against objectivist science and its alleged 'unbiased observations' from a position of nowhere and everywhere. They have rejected this 'nobody's power', as Ermarth (1992:29) calls it, a power that is 'at once human and unspecific, powerfully present but not individualised' because it fails to acknowledge the narrator as an integral part of the story. At the same time, however, there is an acute awareness among feminist environmental theorists of the difficulties that might arise with such sensitivity to the contextuality and constructedness of accounts, since the ability to present an analysis of risks and hazards on such pressing issues as biotechnology, agricultural practices, radiation and the BSE crisis depends fundamentally on reference to a 'reality' outside the constructivity of representation. There is an appreciation, in other words, that acknowledgement of the constructive role played by observer-theorists, their frameworks of understanding and metaphors, creates problems. It makes it difficult, in other words, to speak about theories not being appropriate or adequate to the 'reality' they seek to explicate, to refer to a physical world beyond its description, and to offer a critique of environmental strategies.

This book is an attempt to put into practice and to show the relevance for environmental praxis of some of the insights and commitments arising from the Tutzing 'Time Ecology Project', Gibbons et al.'s (1994) The New Production of Knowledge and feminist theory. This means, in this treatise I fully acknowledge my personal influence on the analysis. I take seriously the demand that theory must not lose touch with experience and everyday practice and, finally, I embrace the recognition that theory is political in the sense that it is the basis for value-laden action. This is not a new development in my writing. Rather, my work to date has been a steady progression towards the theoretical position pursued and promoted here. Thus, for example, my Time and Social Theory (1990) is transdisciplinary in both content and approach. It is programmatic to the extent that it argues that social scientists have to engage with 'nature', technology and the work of natural scientists if their work is to be relevant to some of the key social issues of the late twentieth century such as globalisation and the impact of technology on the environment. In Timewatch (1995a) I once more use the focus on time to effect a shift in basic assumptions, largely in the social sciences but also, implicitly, in everyday life. This is achieved through a substantive focus on health, education, work, globalisation and environmental pollution. In both books, theory is understood with reference to action; it is conceived as praxis. In both texts I have steered a path that avoids the unacceptable choices of traditional social theory and analysis: between biological and social determinism (where people are understood to be governed by either their biology or society), between realism and relativism (where the external world is thought to be either discovered or constructed by the understanding we bring to it), between meta-narratives and particularism (where analyses are considered to be embedded in the worlds of either overarching, universal theories or particular, unique contexts and events). I have shown how to take account of nature without succumbing to biological determinism, how to accept relativism as inescapable without losing the ability to talk about the physical world of 'nature' and technology, and how to be critical of metanarratives in a social world of global/ising relations.

With this book, I feel, there is a need to take this process one step further with reference to both the practice-oriented conceptual approach and its addressees. To this end I gather information from an eclectic range of sources. I engage with phenomena and processes marked by in/visibility, im/materiality, futurity and un/certainty, and demystify the capacity of science to provide truth(s). I tell stories that make taken-for-granted assumptions visible and attend to previously disattended ways of seeing. This entails that I treat implicit theories and conceptualisations, assumptions and presuppositions not as 'second order questions' (Benton and Redclift, 1994:2) but as primary data, as sources equivalent to any other empirical data which social scientists might investigate. It means further that I am in pursuit of theory that is generated from a contextual, 'earthed', embodied position that reflectively acknowledges the theorist as part of her story and analysis. The focus on time enormously aids this endeavour. Since time permeates every aspect of existence, it functions as a constant reminder to the physicality of my being, that I am an embodied person inescapably implicated in my subject matter.

I illuminate events in the round, that is, from a multitude of angles, leaving readers to make their own judgements about 'right' and 'wrong', 'good' and 'evil'. More crucially, I am committed to creating theory that is meaningful not only to social scientists but also to people as they are affected by global/local (from now on *glocal*) socio-economic relations and the environmental impact of the industrial way of life. Currently, such people are outside the stakeholder circle of this social science discourse. This gap between social science explanation and lay perception of environmental issues, however, is not the focus of my attention. Instead, I consider that gap as something to be narrowed. As cultural theorists such as Jacques Derrida and his followers produce ever less accessible theory I want to move in the opposite direction: the more complex the ideas to be presented, the more accessible I feel needs to be the form in which the story is told.

I see this move towards theoretical transparency and public accessibility as essential for a number of reasons. First, in modern industrial societies the primary and most widely accessible 'social theory' is provided by the media, that is, press, radio and television. Journalists, however, do not consider the production of high quality social theory and analysis as one of their primary tasks. In the tabloid press this function of the media is fulfilled particularly poorly. Sadly, therefore, the quality of media social theory stands in an inverse relationship to the number of readers it reaches. As I show in Chapter 5, the public are served badly in cases where social analysis is left to the theoretical and analytical prowess of journalists. In the light of this worrying state of affairs, the approach presented here is intended as an essential supplement to the primary social theories of contemporary industrial society. The objective, however, is not a critique of the sort of social theory offered by the media; it is not a mere academic exercise. Rather, it is to focus on the 'parts' of socio-environmental life that media social theory cannot reach. I recognise that it might be impossible to bridge the gap between academic social theory and the media theory provided by tabloid press. But I feel confident that the issues I raise in this book connect with the experiences and concerns of recipients of media analyses in the widest sense.

A second reason for the pursuit of theoretical accessibility relates to the nature of current environmental issues. Where in previous historical periods academic work and scientific research would be left to trickle through the social fabric unaided and at its own pace, the compression and concentration of environmental impacts of the industrial way of life mean that the traditional path of communication between academia and public life has become too slow. With the acceleration of technological innovation and its intensifying effects world-wide on socio-environmental life, there is now a need to find more direct paths of communication. Speedy and direct multiplex communications seem to be of the essence when inappropriate habits of mind today guide actions that create at an ever-increasing rate long-term, time–space distantiated hazards for generations hence. The days of the trickling-through mode of academic (social) science are numbered. The search for new public theory and institutional structures is on the agenda. The focus on time is central to this endeavour.

### A brief note on 'we'

In this book I tend to use the 'we' as my mode of address to the reader. But who is included in this 'we'? This is an important question. In a book that argues the futility of the quest for objectivity and static truth it makes sense to use the 'we' in an unproblematic way. If I acknowledge and argue that what I can see and understand depends on and changes with context, then the 'we' cannot be fixed either. It too changes with the context.

In addition to the diversity of my social self, therefore, the 'we' encompasses the readers of this book; fellow social/environmental theorists; (social) science colleagues; other women, mothers, daughters, wives, and parents; fellow Europeans and people sharing the same historical period; members of industrial societies; people who have grown up with Western philosophic and scientific traditions; other human beings as earth dwellers who share their habitat with all other life forms at this or other specified times. There are no boundaries to this 'we'. It is indefinitely extendible in time and space. The context, as I have noted above, determines which kind of 'we' is evoked. Only where there is ambiguity and possible doubt, therefore, will I specify the particular inclusiveness of the 'we'.

## Time for the environment

Put at its simplest, the argument of this book is as follows: steeped in the thought traditions of the industrialised West, we learn about and relate knowledgeably to a multidimensional space, but our understanding of the temporal dimension of socio-environmental life is pretty much exhausted with knowledge about the time of calendars and clocks. Nature, the environment and sustainability, however, are not merely matters of space but fundamentally temporal realms, processes and concepts. Their temporality, furthermore, is far from simple and singular. It is multi-dimensional, a multiplex aspect of earthly existence. Without a deep knowledge of this temporal complexity, I suggest, environmental action and policy is bound to run aground, unable to lift itself from the spatial dead-end of its own making.

The prevailing knowledge of clock time, moreover, is intimately tied to the conceptual principles of Newtonian physics and the linear perspective, which encompass within their knowledge frame assumptions about linear causality on the one hand and reversibility on the other, as well as abstraction, rationalisation and objectivity. These assumptions have material consequences which stand in a problematic relationship to the contextual, irreversible temporalities of life and the multiple rhythmicities of nature. The conflict is twofold: first, this industrial time is centrally implicated in the construction of environmental degradation and hazards; second, as a panacea it worsens the damage. Industrial time, in other words, is both part of the problem and applied as a solution. As long as time is taken for granted as the mere framework within which action takes place and is used in a pre-conscious, pre-theoretical way, I consequently argue, it will continue to form a central part of the deep structure of environmental damage wrought by the industrial way of life.

In this book I therefore explore the timescapes associated with that way of life. This entails concern with approaches to time and the multiple intersections of the times of culture and the socio-physical environment. The aim is to make the taken-for-granted visible, to render explicit as well as question what is currently assumed 'natural'. To this end I focus on the conflicts that arise within the industrial modes of life from a) the complexity and inter-penetrations of rhythms: cosmic, natural and cultural; b) the imposition of industrial time on the rhythmicity and pace of ecosystems; and c) the prevailing emphasis on visible materiality and quantity at the expense of that which is hidden from view and latent. Such a time-based analysis of contemporary socio-environmental phenomena and processes not merely shifts traditional emphases and dislodges taken-for-granted assumptions, but allows, in addition, for an innovative revision of approaches and strategies associated with some of the most intractable environmental hazards to date.

While space is associated with visible matter and sense data, time is the invisible 'other', that which works outside and beyond the reach of our senses. This makes time such a pertinent focus for environmental issues. Whether we are encountering chemical processes, ozone depletion, air and water pollution, radiation, or a new disease such as BSE, we are dealing with phenomena where the impacts of actions work invisibly below the surface until they materialise as symptoms – some time, somewhere. At the point of materialisation, however, they are no longer traceable with certainty to original sources. That is to say, these industrially produced phenomena and processes are characterised by invisibility and periods of latency. More often than not, they are recognisable only once they have been identified through the mediating loop of science and technology. This means, many of the products of the industrial way of life are not graspable with the conceptual tools of their construction. In order to engage successfully at the level of practice with the dangers we cannot see, hear, taste, touch or smell, therefore, new theoretical tools and strategies for sustainable action are needed.

Nuclear radiation can serve as an initial illustration. (See Chapter 7 for an extensive treatment of the temporalities of radiation.) Radiation works silently and invisibly from within. It is known only to our cells. As such, it proceeds outside the everyday reach of our senses. Its materiality extending beyond the capacity of human perception and sensibility (except where extended by scientific instruments), radiation affects the collective present and long-term future, our own and other species' daughters and sons of a thousand years hence. It permeates all life forms to varying degrees and it disregards boundaries: skin, clothes and walls, cities and nations, the demarcation between the elements. It is a fate that we share with a global community of beings. Unbounded, it is dispersed in time and space and marked by complex temporalities and time–space configurations. Its life cycles of decay span from nanoseconds to millennia. This means its time horizon too exceeds human capability and concern. Thus, at the level of everyday life, the 'materiality' of radiation falls outside the conventional definition of 'the real', outside conceptions where real means material and where this in turn is defined by its accessibility to the senses. Invisibility, vast, incredibly fast, and variable time-spans of decay, networked interdependence and the fact that effects are not tied to the time and place of emission, therefore, make radiation a cultural phenomenon that poses problems for traditional ways of knowing and relating to the material world. As such, radiation is one of the prime examples of contemporary phenomena and processes whose temporality extend beyond industrial relations and approaches to time; others include chemical and petrochemical pollution, global warming and the depletion of ozone, the effects of genetically modified organisms (GMOs) and culturally induced diseases such as BSE.

My attempt to bring the complexity of time(s) to the forefront of attention allows me at the same time to bring into view the in/visible, latent, immanent and implicate dimensions of socio-environmental phenomena and processes. Such an approach almost by default facilitates the creation of new conceptual tools for practical engagement with the less tangible phenomena and processes of the glocal environment mentioned above. In the light of their diverse temporalities, 'nature', 'culture' and 'the environment' are reconceived. On the basis of concern with matters of environmental time–space, I am able to revise views on the role of money and markets, mediated knowledge and the manufacture of uncertainty, values and responsibility. And, finally, my effort to shift the emphasis in environmental praxis from explicit space and implicit time to the complex temporalities of contextual being, becoming and dwelling has led to the development of the notion of timescapes. To understand socio-environmental phenomena and processes in terms of timescapes is a way of seeing and a conceptual approach that permeates and informs the substance of this book. Where other scapes such as landscapes, cityscapes and seascapes mark the spatial features of past and present activities and interactions of organisms and matter, timescapes emphasise their rhythmicities, their timings and tempos, their changes and contingencies. A timescape perspective stresses the temporal features of living. Through timescapes, contextual temporal practices become tangible. Timescapes are thus the embodiment of practiced approaches to time. Such an understanding, I want to suggest, has tremendous benefits for environmental praxis and the potential to create sustainable futures.

## Timescapes of Modernity: an overview of the chapters

Conflicts between timescapes permeate these pages with each chapter focused on different aspects of created stresses as they arise from the socio-environmental impacts of industrialisation and its specific configurations of time. Industrial time, I propose, is centrally structured to a) the invariable beat of the clock, b) the economic commodification of time and c) the scientific use of time as measure of abstract motion. In this triple configuration, industrial time is central to the discussions on environmental economics and politics as well as the way the media present environmental issues. It is deeply implicated in scientific and political approaches to nuclear and gene technologies. It is centrally tied to the hazards associated with industrial food production and agricultural practices. And, finally, it is crucially involved in approaches to nature, sustainable development and the construction of socio-environmental futures.

Together, these three aspects of industrial time – machine, economic and laboratory time – form a powerful conceptual bloc: time becomes a quantifiable resource that is open to manipulation, management and control, and subject to commodification, allocation, use and abuse. Emphasis is placed on

11

visible materiality at the expense of that which is latent, immanent and hidden from view: the bulk below the surface remains inaccessible. The complex temporalities of the majority of environmental degradations and hazards, however, are located outside the reach of this particular conception. That is, a large proportion of the processes associated with the most difficult environmental problems tend to be inaccessible to the senses, invisible until they materialise as symptoms.

In the first two chapters I set out the prevailing scientific and economic habits of mind and explore their impact on environmental action. In the chapters that follow, those habits of mind are displayed in their institutional and socio-environmental effects: that is, in their permeation of environmental politics and policy, agricultural practice and food production; 'green' technology, scientific innovation and media coverage; as well as environmental activism on the one hand and environmental degradation and hazards production on the other.

The habits of mind under scrutiny in the first chapter relate to the tradition of (Western) industrial societies that understands nature in dualistic terms as the 'other' of culture, as that which it is not: artefacts, culture, Self, humans and the cultivated realm of agriculture. As countryside and meadows, mountains and forests, wild animals and birds, this vision refers exclusively to the *products* of nature, to the externalised outcomes of processes, to de-contextualised physical phenomena without activity and process. As a living entity, however, nature is active and changing and its processes are contingent upon contexts: birds are nesting and migrating at specific times and places; a localised countryside is changing colour with the seasons; specific mountains are showing signs of erosion. That is to say, temporality and context are essential to life and thus to any representation of living phenomena and processes. Without the contextual time–space of activity nature remains abstract and remote, detached from Self, cultural activity and humanity. Thus, traditional habits of mind with their exclusive focus on nature as product, as external framework for human activity and as economic resource fail to take account of the immanent forces that give rise to the phenomena identified with nature. They exclude the energy as well as the re/productive and re/generative capacities of nature that operate irrespective and despite of human activity: the sprouting of new growth after a tree has been felled, the mutations which emerge in response to herbicides and pesticides, in other words, that which humans battle against and seek to bring under their control. As this force works below the surface and beyond the reach of our senses it tends to fall outside the remit of scientific investigation and measurement and, not surprisingly, therefore, be neglected by both physical and social sciences.

When we focus instead on the timescapes of socio-environmental phenomena and events, as I show in the first chapter, a very different picture emerges. First, we recognise that the spheres of nature and culture are not as neatly separable as common language use would lead us to suspect. Second, the centrality of earthly rhythms comes to the fore. Third, the pervasiveness of Newtonian physics in socio-environmental and everyday theory becomes apparent and the positive as well as negative power of the Newtonian habits of mind visible. Finally, the trust in science and its capacity for prediction and control is severely tempered.

Chapter 1 therefore points out that seasons and tidal extremes, for example, are affected by industrial activity just as some of the limits to these activities are set by the fact that humans are tied to the rhythms of night and day – that we, alongside most other living beings, are constituted by a multitude of circa rhythms. These rhythms range from the very fast firing of neurones to the heart-beat, from digestive to activity-and-rest cycles, and from the menstrual cycle to the larger regenerative processes of growth and decay, birth and death. Those internal and species-specific rhythms, moreover, pulse in synchrony with the rhythms of the cosmos. Environmental changes from dark to light, warm to cold, wet to dry set the developmental pattern for all life on this planet, to be internalised and adapted to for specific evolutionary and environmental niches. From cells to organs and even brain activity, our physiology is tied to those periodicities. Women's reproductive cycles are tuned to it and so are our collective activity and rest patterns – all superbly timed and orchestrated into a symphony of rhythms. Sickness and even deaths tend to cluster around specific times of the day, and be synchronised with the temporal patterning of our earth: asthma attacks shortly after midnight, heart attacks and strokes around nine o'clock in the morning, onset of fever from bacterial infection between early morning and midday, and fever onset from viral infections between early afternoon and evening (Rose, 1989:87–90).

This multitude of coordinated environmental and internal rhythms gives a dynamic structure to our lives that permeates every level and facet of our existence. It constitutes temporal frameworks within which activities are not only organised and planned but also timed and synchronised at varying speeds and intensity, and orchestrated to intricate scores of beginnings and ends, sequences, durations and pauses. All aspects interpenetrate and have a bearing on each other. All coexist and are lived simultaneously. All are known at an everyday level with varying degrees of clarity from the taken-for-granted to the theorised. A symphony of rhythms and temporalities thus underpins our development as humans and as living organisms. It marks us as creatures of this earth, as beings that are constituted by a double temporality: rhythmically structured within and embedded in the rhythmic organisation of the cosmos. From a temporal perspective there is no nature–culture duality: we are nature, we constitute nature and we create nature through our actions in conditions that are largely pre-set for us by evolution and history. Instead of emphasis on dualities – such as external and internal, spatial and temporal, natural and cultural – focus on time(s) facilitates additional understanding of the interactive and constitutive aspects of socio-environmental praxis.

And yet, simultaneous with the transcendence of dualisms we are forced to recognise important distinctions between cultural time(s) and the temporalities of nature while appreciating their mutual interpenetration and influence. In the case of industrial societies, cultural time(s) are predominantly rooted, as I suggested above, in the habits of mind associated with technology, classical economics and science. The interaction and conflictual interdependence of these divergent timescapes are theorised with reference to environmental degradation and the creation of socio-environmental hazards.

In Chapter 2 the economic relation to time and its role in both the promotion of 'green business' and the production of environmental hazards are scrutinised. Thus, in the light of the overall argument about timescapes, the focus is on a number of key assumptions and practices associated with neo/classical economics and industrial time: the valorisation of calculation and its associated dependence on quantification, measurement and linear causality, the reliance on visible surface phenomena – the *Merkwelt* and *natura naturata* – on past-based knowledge and on the pursuit of certainty. Cost-Benefit Analysis (CBA) is used as exemplar for neo/classical economic approaches that depend on a combination of the above for their taken-for-granted assumptions, and demonstrates their shortcomings for economic environmental (from now on economental) management, control and risk assessment. The crippling effects of the practice of discounting the future are exposed, and its impact on our capacity to take account of the long-term future on the one hand and the potential to safeguard sustainability on the other is considered. Finally, the temporality of economic growth is distinguished from the timescape of living processes of re/production and re/generation.

The time of economic exchange functions as an abstract exchange value, it translates the work of people and machines into money. As such, it depends centrally on quantification which is achievable only on the basis of the rationalised and decontextualised time of the clock. Clock time is based on the principle of repetition without change. Distanced from the variable rhythms and contextual differences of living systems it recasts time in an atemporal form. As such it can be applied anywhere and any time. In contrast to commodified time, however, the rhythmically constituted processes of ecological transactions and reproduction are not easily quantified and decontextualised. This makes their translation into money almost impossible. In a world where money is synonymous with power, any time that cannot be given a money value is by definition associated with a lack of power and falls outside the value system of economic relations of production and consumption. The time of ecological give-and-take becomes subsumed under the time logic of economic exchange, consumption and globalised market forces. The result is out-of-sync time frames, a prominent feature of a great number of environmental problems: ground water, top soil and forest eco-systems that took thousands of years to develop are exploited in centuries and decades. The time-scale of their reproduction thus stands in an inverse relation to the time-scale of their use, degradation, depletion and destruction. I focus on these and other distinctions not in order to establish a new dualism of natural and industrial commodified time, not to erect a hierarchy of values, not to promote the authority of one over the other. Instead, I elaborate those temporalities in order to bring to the fore the conflicts that arise from their intersections and to provide a base from which to make conscious and informed choices for action.

Throughout this chapter I show the inappropriateness of neo/classical economic assumptions not just for dealing with environmental hazards marked by latency and invisibility but for the establishment of a sustainable, time–space distantiated economy. The prognosis for the future, however, is not as bleak as the analysis would suggest: the necessary leap of the imagination is both huge and minimal. It is radical and dramatic at the level of unquestioned habits of mind that have informed the rise of industrialisation and the global economy to the present and, at the same time, unspectacular and easy since it is lived already in the moonlight economy of environmental praxis: in the household and the garden, in demonstrations for animal rights and the concerns associated with environmental protests. Where the official economy with its principle assumptions of objectivity and abstract self-interest fails, the informal, ethical and temporally sensitive *oikos* economy is eminently suited to step into the breach.

Chapter 3 shifts the focus of attention to liberal democratic political practice and explores the role of the habits of mind, discussed in Chapters 1 and 2, for environmental politics. The material effects of these unquestioned assumptions are theorised with reference to the historically rooted ideal of democracy on the one hand and globalised economics on the other. The chapter demonstrates how central decisions with extremely long-term socio-environmental effects are abdicated to science and transnational corporations, neither of which are socially or politically accountable for their deeds. It suggests that the situation is further exacerbated by the Liberal Democracies' unequivocal commitment to a global economy and the quest for sustained economic growth. With their economistic perspective which tends to elide the distinction between rights of people and rights of money, Liberal Democracies have difficulty taking account of the needs and/or rights of anything or any being without economic power or political influence. Future people and environmental sustainability, therefore, are not readily encompassed within the boundaries of their concerns. Finally, the chapter outlines the disjuncture between the spatially oriented politics of nation states and the complex temporal features of socio-environmental hazards, and identifies the need for a temporalising of democracy which could close the credibility gap between current liberal democratic practice and the popularly held democratic ideal.

In Chapter 4 I utilise the focus on time to bring to the fore difficulties which arise when the rhythmic organisation and time scales of nature are denied or ignored and when cultural constructions which work on the basis of different temporal principles are superimposed as alternatives not just on the everyday lives of humans but on the livestocks and crops associated with agricultural production. Industrial agriculture with its dependence on science, technology and global economics, and its allied emphasis on the times of science, machines and economic relations, is a pertinent case in point. The ensuing clash of principles between these divergent temporal systems – industrial as opposed to the rhythmicity of life and ecological relations – means that their interactions and interpenetrations entail costs and losses that feed into environmental crises. A brief look at nature in the laboratory and laboratory time can illustrate the wider argument.

Since science predominantly studies nature in the laboratory, the subject matter of science is invariably severed from its networked ecological context and the rhythmicity of life. That is, laboratory nature is abstracted from its temporal interconnections and contextual dependencies. In laboratory science, therefore, rhythmic interdependencies are negated and the contextual, embedded temporality of living beings (seemingly) becomes an irrelevance (on 'laboratory time', see also Nowotny, 1994, Chapter 4). A number of implications follow from this move. First, abstracted from interdependencies and context, processes can be controlled, programmed, manipulated, changed, speeded up and slowed down. Second, everything is available at any time and in readiness for use 24 hours a day, 365 days a year. This control of time and constant availability of products finds everyday expression in the arhythmic and decontextualised non-stop principle, just-in-time production processes and the consumer expectancy of being able to buy seasonal foods everywhere and at all times. The taking for granted of such transcendence of rhythmicity and seasonality, I show in this chapter, has dramatic effects on the health and safety of people and the environment.

Once upon a time and distant places it was and is considered 'natural' for the seasons, climate, weather and location to impose limits on human activities, just as it was/is taken-for-granted that these restrictions pose a challenge to human ingenuity. Despite their quest to overcome the vagaries of the weather and to transcend the climatic extremes of the seasons, people were/are embedded in the light and dark, wet and dry, cold and warm, growth and decay, birth and death cycles of nature's earthly rhythms. As long as the human production of food remains seasonal, which also means predominantly local or at least regional, and as long as the primary producers of food retain control over the means not just of production but reproduction, the system remains one of contextual, embedded, interdependent growth cycles. With new methods of food production, processing and preservation, this dependence on time and space has been largely transcended. With globally sourced foods the relative monotony of seasonal produce is replaced by the absolute monotony of the same chemically assisted (and, more recently, irradiated) jet-setting foods that are available to citizens of industrial/ising societies everywhere and all the time.

Such aseasonality and decontextualisation are achieved at a price. They are accomplished at the expense of the health and well-being of citizens, of farmers, their livestocks and their land, and of the wider globally networked environment. Thus, contemporary food hazards are intimately tied to this transcendence of time and space associated with industrial methods of food production and preservation. Focus on the temporal dimension of these production processes shows hitherto disattended relationships and makes hidden dangers visible and tangible. It makes explicit what so far had been implicit connections between science, the chemical industry and agricultural production.

Finally, this chapter considers the conflicting concerns and risk strategies of the major actors in the food system − farmers, traders and food processing corporations − and shows how farmers occupy by far the highest risk position in terms of both their livelihood and their health. Focus on the timescape of agricultural practice, food production, the trade in globally sourced foods and its impact on citizens facilitates a novel interpretation of those inter-dependencies and leads to unconventional practical suggestions for change. First, I argue that in order to safeguard a sustainable environmental future, farmers need to reclaim not only the ownership of the means of reproduction but also of time, since this combined ownership restores to them control over the pace of agricultural life and its sustainable future. Second, I propose that locally grown, fresh produce and full information on the life histories of foods need to be granted as basic rights, since this seems to be one of the few sure ways to secure citizens' safety, health and well-being.

In Chapter 5 I discuss the role of the news media as a primary source of public information about environmental hazards. I use the recent tragic developments associated with BSE in UK cattle and Creutzfeldt-Jacob Disease (CJD) to bring to the fore issues that have thus far escaped attention. I use the focus on time to foreground the dissonances that arise for news-workers faced with the task of treating environmental hazards as 'news' and making them conform with such news values as immediacy, novelty, recency, here-and-now, urgency, timeliness and mediagenicity. Clearly, these values which apply to the description of accidents, crimes, disasters and econo-political events are ill-suited to the reporting of environmental hazards. They are largely inappropriate to environmental hazards since these tend to be long-term, chronic and cumulative and because their dangers tend to be largely invisible. Such phenomena do not require factual description of the here and now but, rather, necessitate historically located analysis of knowledge that is inescapably mediated by competing scientific interpret-ations. This demands of newspaper journalists a mode of operating that they are neither equipped nor motivated to achieve.

Instead of rising to this particular challenge, the media in general and newsworkers in particular tend to convert environmental hazards into matters of economic concern. Thus, in the case of the UK BSE crisis, a highly complex health hazard issue is transformed into an economic beef crisis and with it restored to the journalistic world of certainty and facts, of risk calculation and the descriptive reporting of dramatic outcomes: empty cattle markets, figures about lost revenues to the afflicted countries, menus of schools and restaurants, and political statements.

Here as in previous chapters, I show how the habits of mind discussed in Chapters 1 and 2 work against the necessary confrontation and engagement with the new challenges presented by the hazards that accompany the industrial way of life. But, similar to the conclusions in previous chapters, with respect to the media reporting on the environment, there is not merely a need to change habits of mind and journalistic tradition and culture. The problem is more far-reaching than that. With hazards of this kind, there arises the need for new institutional structures that are able to provide public information and analyses outside the high-pressure, economically competitive framework of press and television news.

In Chapter 6 the complex temporality of environmental hazards is given its most extensive treatment. The fundamental multiplicity of times is explored with reference to the nuclear explosion at Chernobyl and its socio-environmental aftermaths. In this chapter I thus complete the task I began in Chapter 1 with the discussion of scientific habits of mind and their seepage into everyday praxis to a point where today they are taken for granted, naturalised as common sense. Chernobyl brought to the fore the contradictions and incompatibilities of those presuppositions that are rooted in classical science and the linear perspective. As such it pointed to the mismatch between the perspective and the phenomena to which it was applied. Serious engagement at the conceptual level forces analysts to come to terms with complexity, simultaneity and the multiplicity of times.

In the last chapter I use the focus on genotechnology to revisit the key issues that arose in the course of the investigation, not to summarise and conclude but to open up the debate and create a sense of dis-ease and discomfort befitting the contemporary condition. With genotechnology, it needs to be appreciated, the hazard potential of the industrial way of life comes to a head. The temporal features are amplified: time–space distantiation and latency periods are vastly expanded, the link between cause and effect ever more tenuous. The time span of action is compressed whilst the horizon of effects reaches to the end of time. Moreover, the impact of genotechnology on socio-environmental existence is extensive and comprehensive. Genotechnology affects glocal public health, agriculture and food production, as well as trade, export and financial markets and, as patenting and seed monopolies take root, the livelihood of peoples the world over. Like previous technologies it promises cornucopia and an end to the problems associated with the industrial way of life. From a timescape perspective, however, we can see that the promises of success and unprecedented financial rewards are squarely matched by the technology's hazard potential. Equally, the scientists' control over laboratory processes is matched by the absolute loss of control over the 'products' once they have entered their new environments as free agents. 'Uncertainty and risk', as Regine Kollek (1995a:103) argues, is 'the price which must be paid for the total accessibility and control of these living objects in the laboratory.' Not only can these living things not be recalled but they cannot even be traced as they migrate and mutate through the networked earth system of ecological relations.

The timescape perspective enables us to see the invisible. It facilitates recognition that the hazard potential of the industrial way of life is inescapably and proportionally related to its successes – the global dispersal and reach of its technological innovations, the transformation of cyclical reproduction processes into linear energy-input waste-output production systems, the transcendence of seasonal, climatic and regional dependence, the promise of unlimited supplies of food. As such it helps not only to reformulate environmental problems and their cures but also to re-articulate the interpenetration of the spheres of safety and certainty, risks and hazards, sustainable action and precautionary principles. Rethinking environmental issues in temporal terms thus gives us considerable theoretical and practical access to their complexity, which in turn provides the potential for alternative socio-environmental praxis.

# Part I

# HABITS OF THE MIND
Environmental timescapes conceptualised

# 1

# NATURE RE/CONSTITUTED AND RE/CONCEPTUALISED

## Mapping the scope of industrial traditions of thought

## Introduction

Contemporary environmental hazards make it difficult to conceive of nature and culture as separate. Whether we think of the ozone hole, pollution, or food scares, nature is inescapably contaminated by human activity, that is, by a way of life that is practised and exported by industrial societies. This loss of clear distinction between nature and culture, however, is not only brought about by the industrialisation of nature but also by the hazards that endanger humans, animals and plants alike. That is to say, the common danger has a levelling effect that whittles away some of the painstakingly erected boundaries. It erodes the meticulously constructed and carefully guarded differences between humans and (the rest of) nature, between creators of culture and creatures of instinct, or to use an earlier distinction, between beings with and beings without a soul.

> In the threat people have the experience that they breathe like plants, and live *from* water as the fish live *in* water. The toxic threat makes them sense that they participate with their bodies in things – 'a metabolic process with consciousness and morality' – and consequently, that they can be eroded like the stones and the trees in the acid rain.
>
> (Beck, 1992a:74)

Of course, the distinction between nature and culture could be upheld only on the basis of a strict separation of mind and body and the consignment of the latter to a lower level of existence. Whilst bodies and the dependence on birth and death remind humans of the inescapability of the earthly dimension of their being, cultural activities facilitate the belief in difference and a distancing from all things natural. Through mind-based culture, humans have sought to transcend their earth-bound condition and the limits set by nature: art and writing, ritual and exchange based on money are all ways to overcome earthly transience and human finitude. Technology has played a particularly pertinent role in this flight from nature. Technologies ranging from the axe to nuclear power are expressions of the human effort to extend the 'natural' powers of the body and the senses by artificial means: x-rays to extend vision, wheels and later machines to speed up movement across space, computers to expand the capacity of the brain, genotechnology to redesign and compress evolutionary processes. Collectively, these developments in technology have contributed to the ever-increasing distance between natural processes and cultural activity. In the light of this history, it is not surprising that (Western) industrial societies' intellectual history is marked by an effort to put on ever firmer footings the difference between human culture and nature, as well as the distance between human mental activity and the physicality of being.

Some of the successes of this progressive dissociation, however, have developed into environmental hazards. Radiation from nuclear power, disruption of hormonal function from synthetic chemicals, and pollution from the use of chlorofluorocarbons (CFC's), the burning of fossil fuels and motorised transport are just a few pertinent illustrations of where intended technological developments result in unforeseen, unwanted problems. As existential threats, these industrial achievements bring us once more face to face with our being rooted in and dependent upon nature. As two sides of the same coin, success and excess have brought us full circle to confront the basis of our existence and present us now with the opportunity to re-think the meaning of human being and finitude. This shifting ground of conception and knowledge presents an opening for seeking fresh answers to the question 'How shall we live?' (see also Adam, 1996a:99; Beck, 1992a:28; Giddens, 1991:215, 223; Weber, 1919/1985:143, 152–3).

A first step towards such a re-vision, it seems to me, would be to break down the old boundaries and to reconfigure the nature–culture divide. There is, however, as yet little evidence that this is happening. It seems that the conceptual implications of the environmental threats have not yet penetrated everyday understandings of nature. That is to say, despite the permeation of nature by technology and irrespective of the recognition that environmental hazards do not discriminate between animals and humans, the nature–culture distinction continues unabated as the dominant everyday conception in (Western) industrial societies. In other words, lay understanding, especially among city dwellers, remains unchanged: nature means green fields and pretty countryside, existing 'out there' as a place for leisure and stress relief, aesthetic consumption and redemption.

A second step would be to bring the complex temporality of being to the forefront of understanding of that existential question. At present, a noteworthy feature of contemporary understandings of environmental hazards is their emphasis on space. At one level, of course, this is not surprising given the spatial reach of the problems and their noted disregard for boundaries. What is surprising, however, is the lack of focus on time, considering the striking temporal characteristics of contemporary environmental phenomena ranging from pollution and degradation of the environment to radiation and the disruption of the endocrine systems in all species. Gone is the simple world of linear causal links between input and output, of proportional relations between action and effect, of human time-scales for political planning horizons. Gone too is the comforting dependence on scientific certainty and accurate predictions. The incubation time of Bovine Spongiform Encephalopathy (BSE or mad cow disease), for example, is thought to be somewhere between five and twenty (possibly even fifty) years, whilst scientific predictions of the temporal boundaries of the effects of Chernobyl on Cumbrian sheep have expanded from three weeks to six weeks and more until, ten years on, the scientists have given up guessing. Today, they cannot even be certain that their initial distinction between Sellafield and Chernobyl radiation can be upheld (see Wynne, 1992, 1996). Oil disasters like the one in February 1996 off the coast of West Wales, where the *Sea Empress* discharged 70,000 tonnes of light crude oil into the coastal waters, have had a range of time-based effects: immediate and visible ones as well as quantifiable but open-ended ones, such as the decimation of the sea bird population and the contamination of marine life. By far the largest proportion of consequences, however, are invisible and, by the time they finally materialise as interconnected webs of symptoms, they will not be traceable with certainty to any particular causes. Given such extraordinary temporal features, one would have expected there to be a steeply rising interest in the time dimension of environmental processes among all who are concerned with and affected by such matters. There is, however, as yet no evidence of such change in focus, just as there is no significant shift in perspectives on nature. Instead, emphases remain firmly on space as the physical realm which is accessible to sense data and amenable to quantification. This book is a first sustained attempt to remedy that situation.

In this first chapter, I want to draw out some of the connections between the persistent nature–culture divide, the emphasis on visible spatiality and the Newtonian science tradition of thought, and to show how a timescape perspective can take us in directions and areas of inquiry that spatial analyses can't reach. Having just argued for the need to reassess the nature–culture separation, I propose nevertheless that in this temporally charged context of evaporating distinctions it is important also to understand some of the time-based differences as a precondition to being able to seriously engage in existential questions about ways of life and responsibility. Moreover, focus on the temporalities of the industrial way of life, as distinct from the temporalities and rhythmicities of living nature, is crucial for grasping 'ice-berg phenomena', that is, for understanding hazards that defy the theories and methods of traditional science because they are both visible and invisible, material and immaterial. To express this particular quality of contemporary environmental phenomena, I shall refer to them as in/visible and im/material processes. The foregrounding of such temporal distinctions necessarily entails examination and some renewal of the conceptual tool kit. Thus, to engage with below-the-surface and beyond-the-present issues requires creative conceptual moves beyond the materialist, objectivist praxis of science and, as I argue in the next chapter, innovative socio-economic policies cognisant of the limitations of the Newtonian heritage in both science and economics. It necessitates the unsettling of some habits of mind and the strengthening of others.

## Nature re/constituted

Animals grazing peacefully on a hillside, waves lapping gently up the pebble beach, a pine forest whistling in a storm, a river bursting its banks, a hurricane tossing houses and cars in the air like play-things, a bush fire raging out of control – all are images of nature, some idyllic others threatening. Can we be sure, however, that this is nature in the conventional meaning of the word, that is, the result of forces uncontaminated by human activity and production? What becomes of this understanding of nature when those grazing animals are contaminated with radiation or suffering from BSE, when the waves carry the residues of an oil spill or other forms of pollution, when the pine forest (a monoculture, likely to have been planted during the last century) is suffering from the effects of acid rain, when the flooding is due to agricultural practices that have led to over silting, when the extreme weather conditions are linked to global warming, and when the bush fire and the scale of its damage have been facilitated and exacerbated by human actions? During this century it has become increasingly difficult to sustain the division between nature and culture. When even the stratosphere is affected by the industrial way of life, when the sun is turned from source of health and well-being to health hazard and danger, when the foods we eat are as likely to poison as they are to nourish us, when the water that is to cleanse us is a potential source of skin diseases, when the air we breathe causes respiratory diseases and allergies, when the traditionally conceived untamed, raw power of nature is so extensively influenced by human action then the traditional separation between nature and human culture collapses. In such a context, as Giddens (1991:224) argues, nature is brought 'to an end as a domain external to human knowledge and involvement'.

Recognition of the pervasive impact of human activity on the environment is, of course, not new to this century. During the middle of the last century Karl Marx was reflecting on the acculturation of nature when he wrote in *The German Ideology* (1968:59) that 'the nature which preceded human history no longer exists anywhere'. Already at that time, there seemed few places left on this earth that could be said to be uncontaminated by human cultural activity. The Romantic movement too was responding to the technological encroachment on nature and its exploitation as an inanimate resource. Romantics were writing about a 'return to nature', and representing nature in its wild and untamed forms in their paintings. Towards the end of the nineteenth century, in America, the land of seemingly boundless wilderness, Henry David Thoreau was one of the first writers to identify the potential threats to unadulterated nature. A disciple of the Romantic writer Ralph Waldo Emerson, he proposed that areas of wilderness be preserved, that logging, settling and cultivation be practised in moderation and in harmony with nature. In his books, Thoreau contrasted the secular, materialist and commodified way of life with the natural principle of harmonious cohabitation and its redemptive spirit. The Romantic movement thus tends to be seen as a first response to industrialisation and the loss not just of untamed nature but its spirituality and soul. That is to say, it arose with the commercialisation of nature and the development of a mechanical science that conceived of nature as an inanimate resource for exploration and investigation, human consumption and economic development. Thus, Kate Soper writes,

> Untamed nature begins to figure as a positive and redemptive power only at the point where human mastery over its forces is extensive enough to be experienced as itself a source of danger and alienation. It is only a culture which has begun to register the negative consequences of its industrial achievements that will be inclined to return to the wilderness, or to aestheticise its terrors as a form of foreboding against further advances against its territory.
>
> (Soper, 1995:25)

The increase in environmental movements during the second half of this century further substantiates the thesis that the level of concern with nature and the environment stands in a direct relation to the degree of human alienation and the extent to which nature as uncontaminated nature is rapidly disappearing. In this context, it is the recognition of human endangerment through hazards arising from the industrial way of life that precipitates the increasing interest in nature and environmental issues. Stephan Heiland (1992a, 1992b), however, draws a pertinent distinction between existential threats and threats to the environment. A Nomadic tribe whose basis of existence is threatened by a combination of environmental degradation and 'global development' (see also Leckie, 1995 and *The New Internationalist*, April 1995) is not facing an environmental but an existential crisis. For members of such a tribe the threats constitute an 'in here', not an 'out there' problem in the same way as the beef crisis is conceived by cattle farmers as a threat to their existence and not as an environmental crisis. For the largest proportion of members of industrial societies, in contrast, the problems are still external; the hazards have not yet penetrated to bases of existence. Consequently, issues continue to be conceived in external terms, that is, detached from the Self and projected onto trees and animals, beaches and lakes, heaven and earth. The causes too are understood externally, one step removed from personal actions: chimneys belching out pollution, oil rigs and pipe lines, nuclear power stations and inappropriate animal feed stuffs are just some of the sources associated with out-there environmental hazards which, in turn, threaten us. According to Heiland, therefore, industrial societies are not just the causes of environmental problems they are also the source for conceiving of the dangers in environmental terms. This understanding, in turn, is rooted in a very specific conceptualisation of the nature–culture relation. As this book shows, especially in the first two chapters, it is tied to an emphasis on vision and space, linked to the successes and imperialistic tendencies of the sciences and their products during the past three to four hundred years and connected to structural features of the global economy.

With their intimate connection to power, we can think of the industrial–scientific–economic relation to nature in terms of a 'Faustian bargain'. Rupert Sheldrake (1990:28–9) writes engagingly about this trade of the soul for magical, superhuman powers over life. He, like many others, links the stories of Faust and Frankenstein to the scientific quest for unlimited knowledge and to the products of a knowledge whose un/intended consequences have returned to haunt us. He suggests that members of the British government have talked about their country's nuclear policy in terms of a Faustian bargain from which there is no turning back: a country can stop building nuclear weapons and power stations but the knowledge is irreversible – the genie is out of the bottle, in need of constant control, management and safeguarding. The British nuclear policy and other 'bargains' of this kind were not without consequences; they fundamentally altered the relation between human beings and nature: 'instead of being concerned above all with what nature could do to us', writes Giddens (1994a:102), 'we have now to worry about what we have done to nature.' In the light of this interpenetration and the history of mutual implication of nature and cultural activity, it is interesting to find that the nature–culture distinction in which nature is understood as culture's external other is persisting in both lay and scientific discourse.

## Upholding difference and distance: images of nature as culture's 'other'

Wildlife, animals and plants, countryside and fields, woodlands and mountains, rivers, lakes and the sea, earth, wind and fire, these are some of the most popular images of nature that emerged from my short interviews with young adults. Almost invariably, these images were developed in sharp contrast and great distance to their own city lives. Nature was defined as 'other' – from Self, humans and culture – and away from the hassle and traffic of the city, its smells and built-up areas, its people and its social structures of control. Nature as the countryside and its inhabitants, moreover, was associated with pleasant feelings, a romantic vision of a wholesome world beyond the corrupting influences of the city. As such, my findings overlap with those of Trommer's (1988) investigation of conceptualisations of nature in pupils of a German city school. These young people also thought of nature as external to themselves and associated it with countryside, wildlife and positive, romantic feelings. In both cases, the city-based images of nature have a strangely atemporal, universal and abstract quality. They portray a pleasant but rather unfamiliar, almost alien world of picture postcards and holidays, television programmes and school outings. Finally, Kate Soper, in her book *What is Nature?* summarises lay concepts of nature along very similar lines.

'nature' is used in reference to ordinarily observable features of the world: the 'natural' as opposed to the urban or industrial environment ('landscape', 'wilderness', 'countryside', 'rurality'), animals, domestic and wild, the physical body in space and raw materials. This is the nature of immediate experience and aesthetic appreciation; the nature we have destroyed and polluted and are asked to conserve and preserve.

(Soper, 1995:156)

While my small investigation confirmed the 'ideal types' presented in the literature, it also pointed to an extensive potential and capacity for making distinctions. First, as one might expect, I found a substantial difference between urban and rural conceptions and, second, I discovered marked discriminations being made between images of nature and those of the environment.

I learned that members of rural farming communities credit nature with active powers and with what could almost be described as a 'personality' Nature was widely thought of as something that functions without and despite of human activity: a tree sprouting from the base after it has been cut down, the acorn growing into an oak, the heifer that knows what to do with her newly born calf, the lamb that recognises its mother in a field of other lambs and ewes. There was recognition that humans can destroy parts and aspects of nature, can affect it, manipulate and adapt it to human need and desire, and to some extent even repair it. In the rural conception of nature, however, this influence was distinguished from control since humans were considered to have no control over nature – over tides and seasons, wind and weather, the changing lengths of night and day.

It is interesting to note that such differences in understanding nature have been expressed as early as the Middle Ages when they were formulated as a distinction between *natura naturata* and *natura naturans* (see also Sheldrake, 1990:61 and Chapter 4 more generally). *Natura naturata* was thought of as natured nature, that is, nature's products: mountains, lakes, trees and animals. It is that which we can observe with our senses. *Natura naturans*, in contrast, was understood as the temporally constituted dimension, the force that gives rise to those observable phenomena, the invisible energy that is recognisable only through its products. Even though we might say that we can intuit or 'sense' this force, its processes work unseen below the surface, beyond the reach of our senses. At its inception, *natura naturans* was considered the soul of nature. Rupert Sheldrake explains what happened to *natura naturans* with the development of the physical sciences.

When the founders of mechanistic science expelled souls from nature, leaving only passive matter of motion, they placed all active powers in God. Nature was only *natura naturata*. The invisible productive power, *natura naturans*, was divine rather than physical, supernatural rather than natural.

But this attempt to remove all traces of spontaneous organising activity from nature ran into grave difficulties from the outset. The ghost of the invisible souls remained, in form of invisible forces. Gravitational attraction, acting at a distance, showed that there was more to the physical world than mere passive matter in motion. The nature of light remained mysterious, and so did chemical, electrical and magnetic phenomena.

(Sheldrake, 1990:61–2)

From the rural conceptions by members of the farming community, we can see that the force beyond the senses has been retained as an integral part of understanding. Here, *natura naturans* is that which farmers 'battle' against and engage with on a daily basis. It is also recognised as that which returns to haunt them when successes have turned into excesses, when, for example, ever increasing growth rates and yields have resulted in soil degradation and new diseases of plants, animals and people which are shrouded in mystery, uncertainties, rumours and hearsay. As threats to bases of existence, however, these kind of problems are no longer 'out there' environmental problems. Instead, they are understood in terms of the conflictual interaction with *natura naturans*, as pushing back the boundaries of nature, going against nature by imposing 'unnatural' measures, and nature retaliating.

Despite these substantial differences in rural and urban conceptions of nature, however, I also found a number of shared approaches. In both cases, nature was identified with reference to that which it was not: artefacts, culture, Self, humans and, in the rural case, the cultivated world of agriculture. Nature achieved definitional clarity as 'other', as that which is not created by humans, by being contrasted with the familiar and that which is known. This explains why nature is such a 'slippery' concept with entire books dedicated to its elaboration. Raymond Williams, in *Key Words* (1976:184), goes so far as to designate it 'perhaps the most complex word in the language'. If we think of nature as a dependent variable, however, then the complexity of meanings is no longer a surprise. Understood as 'other' to the familiar world of human production, it clearly *has* to change not just with historical periods but with the multiple contexts in which it is used.

A second shared conception relates to the environment. In contrast to nature as 'other', the environment was defined by the people I talked to in a human-centred way as both the place and surroundings where we live, work and relax. Moreover, it was related not just to where but how we live. Talking of the environment, my respondents from the city took account of the human impact on nature; as such, their imagery was predominantly negative and gloomy. Gone was the idyllic picture of countryside and wildlife, gone the warm feelings. Instead, there were associations with road constructions

31

and pollution, destruction of the countryside and disasters such as oil spills, radiation and floods. Without exception their images of the environment were negatively tied to the human impact on nature. Examples of this human effect on nature extended from the hole in the ozone layer to taking a walk in the countryside. As one person put it,

> Whatever humans are doing in their rural, urban or industrial environments, they have an effect and produce pollution: using fossil fuels, felling trees, having a bath – even the most innocent activity has ultimately a polluting effect.
>
> (T.A.)

In the farming communities this human impact on the environment was conceived more in terms of the hard work of cultivation: maintaining hedges, keeping fields productive, processing effluent. As they saw themselves more intimately and positively engaged at an everyday level with environmental matters, their conception was not predominated by disasters and global hazards.

It is the interactive component in these images and associations that transforms the place where people live into an environment that is fundamentally relative to the context within which it is constructed in mutuality. Thus, the environment could be a bathroom with bather, a gardened garden, or the polluted atmosphere in conjunction with power stations providing energy for homes and work places, driving vehicles emitting exhaust fumes, and cattle farms producing methane. Context, moreover, refers not only to spatial but temporal locations and horizons, the when and over what period of specific actions and processes.

*It is the connection to how we live that turns spaces into timescapes of mutual influence and construction.*

Such environmental time, however, tends not to be made explicit but taken-for-granted as the implicit backdrop to daily life. This 'when' and 'for how long' is the time of history, calendars and clocks, of seasons and diurnal cycles. It provides the external framework within which actions are planned and executed. It is a time that operates independent of human actions, an objective parameter that allows us to locate actions in a temporal grid and consider questions of timing and speed. Beyond this backcloth to phenomena and processes, however, there is an additional aspect of time which is central to conceptions of nature and environment(s). This time is less easily accessible and less tangible than the time of seasons, clocks and calendars. It is a time sensed but rarely talked about, a time that is internal to phenomena and constructed in interaction, a time associated with *natura naturans*.

To explain the difference between these times, let me briefly return to the city dwellers' images of nature, to the mountains, forests, countryside and birds. These visions are references to *natura naturata*, to physical phenomena without activity and process, externalised outcomes of processes that have been de-contextualised and de-temporalised. Temporality and context, however, are essential to any representation of life. That is to say, living nature is active and changing, its processes contingent upon contexts: birds are nesting and migrating at specific times and places; a localised countryside is changing colour with the seasons; named mountains are showing signs of erosion. Without the time–space of activity and processes, that is, nature remains abstract and remote, detached from Self and humanity. The rural images of nature as a force, as *natura naturans*, in contrast, implicitly take into consideration the internal, generative time of phenomena and processes. As the immanent force that transformed our earth from its gaseous state into an environment that evolved life and different species, that turns acorns into oaks and the flap of a butterfly wing into a hurricane, *natura naturans* constitutes temporality. This lived, generative temporality originates in the interactive principle of sociality. I am not here referring merely to human sociality but to the principle of interaction that creates asymmetry and with it difference and change, the precondition for there to be time. Perfect symmetry is an atemporal state that is broken at the point of interaction. Such interaction is the start of sociality, a general principle that covers phenomena from atoms to human beings and their institutions (for a detailed exposition of these thoughts, see Adam, 1990, especially Chapters 1 and 7; Luhmann, 1978, 1982a; Mead, 1939/1952).

To explicate the temporality of nature, therefore, we need to reconnect the externalised phenomena to their generative processes, the countryside to its re/production, the forests to their formation. We need to bring into a conceptual unity *natura naturata* and *natura naturans*, that is, the condition, the medium, and the outcome of interactive processes. We need to see the 'product' produced, nature natured, life lived. The lay conception of the environment, by taking account of interaction and the way we live, is a substantial step in that direction. I want to strengthen this vision by bringing the disattended temporal dimensions to the fore, reconnect what became detached from sense and sensibility. To this end, I want first to look briefly at some of the ways the environment has been theorised, as well as at some of the thought traditions that work powerfully against the temporalisation of understanding, before I offer an initial introductory outline of the disattended knowledge of environmental timescapes in the context of industrial culture.

## The environment: a question of praxis and indeterminate impacts

I use the Marxian concept of praxis as a means to stress the inseparable mutual implication of theory and practice. From the above, we have already seen that the way we conceive of nature and the environment relates to how we approach the world around us and to the fact that the time–space of our lifeworld is implicated in our understanding. This mutuality was theorised during the first part of this century in Germany by Jakob v. Uexküll and Georg Kriszat (1934/1983) who argued that the environment is not a fixed condition but arises from the contextual capacity of a being's consciousness and senses. It means that the same physical space will be a vastly different environment for different species and for members of those species. Thus, the same forest, for example, constitutes a very dissimilar environment for a worm, a tick, a fox cub or its mother, a tourist, a member of a tree-felling crew or a local inhabitant. This theory thus expands on the human centredness of the lay conceptions and confirms the relative, contextual status of 'the environment'. Equally, we could say that the theory is borne out by and makes sense of the difference in image and association between city dwellers and members of urban communities.

The environment, according to v. Uexküll and Kriszat, is composed of both a perception- and an impact-based dimension, the *Merkwelt* and the *Wirkwelt*. The former relates to what we can perceive and notice, that is, take in with the aid of our senses and consciousness, whilst the latter refers to the impact of a being's actions. This distinction can serve as a useful conceptual tool for thinking about highly complex environmental issues. Focus on the 'worlds' of perception and impact bring to the surface some pertinent aspects of contemporary environmental hazards. The first aspect relates to their enormous spatial disjuncture where the *Merkwelt* of perception is always and necessarily contextual and locally constituted. This local contextuality is only extendible in the imagination and with the aid of such technologies as television and computers. The immanent impact of the industrial way of life, the *Wirkwelt*, in contrast, is spatially and temporally open and tends to extend across the globe on the one hand and to the stratosphere and the universe on the other: radiation, synthetic chemicals and genetically engineered organisms being pertinent cases in point. The second aspect brings to light the unbridgeable temporal gap between action and impacts. Many contemporary environmental hazards such as ozone depletion, acid rain, damage to the reproductive and immune system of species or BSE have not arisen as symptoms until years after they began their impact as invisible effects of specific actions. Others only externalise as symptoms after they have combined to form a critical mass. That is to say, the *Wirkwelt* is temporally open and becomes perceivable as *Merkwelt* only after it materialises into a visible phenomenon at some time and some place: pesticides in foods, radiation and chemical damage to the unborn, and global warming are just a few illustrations of this temporal disjuncture. The third point arising from this distinction between *Merk-* and *Wirkwelt*, the worlds of perception and immanent impact, concerns the fundamental perspectivity of beings. This, of course, makes the expectation of certainty and provable knowledge a futile dream.

That is to say, not only is it impossible to know all the networked interconnections that eventually give rise to the symptoms we recognise as such but, even more importantly, the inescapable perspectivity rules such flights of fancy out of the question.

On the basis of this theoretical insight, then, we have no option but to live cautiously and precautionarily, cognisant of the fundamental limits to our contextual, perspectival knowledge and of the time–space indeterminacy of our actions' impacts. As I indicate later in this chapter, however, there are powerful thought traditions and structural features that work against taking account of this indeterminacy. These traditions, as I show in the next chapter, predispose actors towards such measures as environmental impact assessment (EIA) or environmental audits, highly dubious schemes in a context where the worlds of perception and impact very rarely overlap – better than nothing, of course, but in danger of creating a false sense of security.

The recognition of indeterminacy which arises so persuasively from both the temporal focus and the theory of v. Uexküll and Kriszat re-emerges in a different guise from the social theories of Ulrich Beck (1992a, 1994, 1996) and Anthony Giddens (1994a, 1994b) where it is associated with the reflexivity of modernity. Giddens writes of manufactured risks and uncertainties, of conditions for which it is foolhardy to think of risks as mere external side-effects. To Giddens, both risk and uncertainty are the very conditions created and identified by science and technology. Each technological in(ter)vention, once released into the socio-cultural fabric and its environment, has an impact of open-ended duration and scale that is unbounded in time and space. Its un/identified effects, in turn, constitute the conditions for further in(ter)ventions and actions. The understanding of these risks and uncertainties in terms of external side-effects, he suggests, is therefore not only meaningless but, more worryingly, adds to the creation of further hazards. This unending creation of new hazards and manufacture of risks is pertinently critiqued by the work of Colborn *et al.* (1996) who show with reference to synthetic chemicals how each technological 'solution' for technologically caused problems leads to further unpleasant and unanticipated surprises: what were thought to be 'the "safest" chemicals proved to be the most dangerous' (Colborn *et al.*, 1996:244) and what was not known turned out to be more important than what was known.

> Ultimately, the risks that confront us stem from this gap between our technological prowess and our understanding of the systems that support life. We design new technologies at a dizzying pace and deploy them on an unprecedented scale around the world long before we can begin to fathom their possible impact on the global system or ourselves. We have plunged boldly ahead, never acknowledging the dangerous ignorance at the heart of the enterprise. . . .

Our dilemma is like that of a plane hurtling through the fog without a map or instruments.

(Colborn *et al*. 1996:245–6)

A temporal perspective helps to illuminate the nature of the gap between application and long-term effects and between the perception of symptoms (the *Merkwelt*) and the in/visible, in/direct, non/linear, non/proportional impacts of actions (the *Wirkwelt*).

Beck designates contemporary societies as *risk societies*, social organisations for whom risks and environmental hazards instead of goods and wealth become a central structuring feature. These risks, moreover, are no longer calculable and predictable since their latent impacts are unbounded with reference to time and space. Thus, Beck's 'risk society' produces and legitimates hazards that are beyond the control of its institutions: science, politics and the market.

The entry into risk society occurs at the moment when the hazards which are now decided and consequently produced by society *undermine and/or cancel the established safety systems of the provident state's existing risk calculation*. In contrast to early industrial risks, nuclear, chemical, ecological and genetic engineering risks a) can be limited neither in terms of time nor place, b) are not accountable according to established rules of causality, blame and liability, and c) cannot be compensated or insured against. Or, to express it by reference to a single example: the injured of Chernobyl are today, years after the catastrophe, not even all *born* yet.

(Beck, 1996:31)

The indeterminacies associated with contemporary environmental hazards, therefore, are of an ontological-structural and epistemological-cosmological nature. Their reach into industrial societies' knowledge bases is far deeper and their permeation of that social fabric much more extensive than notions of uncertainty, risk and unintended consequences would lead us to believe (see also Szerszynski, Lash and Wynne, 1996; Wynne, 1992, 1996). Thus, emphasis on uncertainty, for example, stresses the temporary nature of not knowing or not knowing to any degree of certainty. That is to say, something is uncertain until further research and scientific investigation are able to provide the expected clarity and certainty. The idea of unintended consequences accentuates rational choice in the conduct of individuals: people make rational choices but not all the consequences are within their controlling grasp. The notion of risk, finally, implies the potential for decisions and calculation. This in turn relies on the dependence that the past is a reliable guide to possible future states. In contrast to these three conceptions associated with the unknown, this book presents the case for irreducible indeterminacy as a condition that has to be encompassed and embraced in personal and public responses to the sort of hazards I engage with in this treatise.

It is interesting and important to note, therefore, that the language of invisible, immanent hazards, manufactured uncertainty and indeterminacy stands in stark contrast to the current public environmental discourses of scientific proof, certainty, prediction of the future based on knowledge of the past, risk calculation, and 'safety in the normal sense of the word' – a term that gained great popularity with the British Government during the BSE crisis. Equally noteworthy is the marked difference between, on the one hand, the invisibility, immanence and latency of many environmental hazards and, on the other, the dependence on sense data and measurement, the assumed homology between materiality and the real, and the almost exclusive focus on space. While it seems strange to continue to depend on such an inappropriate conceptual package, it is important to understand that the historical roots of this understanding go deep and are firmly secured in a number of mutually supporting theoretical traditions: the dualistic understanding of nature and culture outlined above, the linear perspective which I am going to focus on next, and the all-pervasive heritage of Newtonian science which will feature prominently throughout this book and has its contours sketched out later in this chapter.

## Denatured nature: corporeal and visual

The preoccupation with visibility, materiality and space is deeply embedded in the knowledge traditions of Western societies and, as such, this emphasis on physicality is implicated in the everyday understanding of nature and the environment. Romanyshyn (1989), in an outstanding book on technology, takes the invention of the linear-perspective vision as one of the crucial turning points in the development towards the modern techno-scientific way of conceiving the world.

> The invention of linear perspective space initiated a revolution in human life. In the space opened up between the distance point and the vanishing point a new self, a new body, and a new world were born. We are the heirs of that revolution. . . the self as a spectator behind the window has become the world's measure by making the world a matter of vision, an infinite vision which, in its singular focus and fixed intensity, has clarified the mystery of the world's depth in its explanations.
>
> (Romanyshyn, 1989:65)

He argues that this artistic innovation of the fifteenth century became a cultural habit of mind that separates observers from their subject matters, isolates the objects of vision from their context and, finally, fragments those objects behind a mathematical grid. By looking at the world through such a grid, 'reality' is translated from a living temporal process into a set of fixed numerical relations (see also Arendt, 1958, especially pages 257–68). Moreover, the linear-perspective vision shifts the artist from the participatory centre to a position of external, ex-terrestrial spectator, from implicated participant to objective observer. The self behind the mathematical 'window' is a person that casts no shadow. Disembodied, de-temporalised and stripped of feelings and emotions, the living, interactive self is transformed into an eye of distance whose fixed, singular, atemporal viewpoint and neutral, impartial gaze leave its subject matter untouched. The embodied person gets displaced by the head and the mind's eye. The body is left behind, rendered irrelevant to understanding. Matter and vision are conjoined to conquer the sphere of knowledge. They become the modern view on the world.

This means that almost one century before Descartes' work separates mind from body and self from the physical world, this artistic innovation prepared the ground for subsequent philosophical and scientific developments. When nature is understood scientifically today, it is perceived from a position outside the earth by a being that is capable of 'being everywhere at once and nowhere in particular . . . a consciousness at once immobile and omnipresent' (Ingold, 1993:155). Scientific objectivity means just that: it suggests that the position of 'observers' is irrelevant to what they see, that the object of observation is the same irrespective of context, that difference in time and space do not affect it in any way.

> The linear perspective is a celebration of the eye of distance, a created convention which not only extends and elaborates the natural power of vision to survey things from afar, but also elevates that power into a method, a way of knowing, which has defined for us the world with which we are so readily familiar. It is the transformation of the eye into a technology and a redefinition of the world to suit the eye, a world of maps and charts, blueprints and diagrams, the world in which we are, among other things, silent readers of the printed word and users of the camera, the world, finally, in which we have all become astronauts.
>
> (Romanyshyn, 1989:33)

This artistic tool is thus an important precondition for scientific cosmology, experiment and explanation. Moreover, it has become naturalised in everyday understanding. It permeates today the assumptions not only of politicians and journalists of Western industrialised societies but of all their members as they go about their everyday business. It is implicated in conceptions of nature and the environment, in understanding the human–nature and culture–nature relationship, in the development and use of technology, in economic exchanges and in approaches to environmental hazards. It under-pins the preference for space over time and the association of 'the real' with visibility.

It is important for environmental analyses that we recognise this shift in cosmology to be a conceptual move only. We need to appreciate it as a heuristic that transforms an ephemeral world of seething quality and infinite complexity into a static, bounded, quantifiable object for comprehension and study, as a way of changing contingent, interconnected processes into abstracted, isolated parts, and as a means of turning implicated participants whose own being/becoming is inseparably bound up with their environments into spectators that cast no shadows. It is equally important to appreciate that, here as with any other conceptual scheme, whatever does not fit the 'vision' is destined to vanish. Starved of its basis of existence, whatever falls outside the frame of reference ceases to exist for the purpose of understanding and action. This artistic move towards simplification, reduction and exclu-sion has been further developed and perfected in Newtonian physics where it is applied to the understanding and explanation of the material world. That is to say, the central characteristics of the linear perspective vision – abstrac-tion from context, objective observation, quantification of sense data and the single fixed focus – constitute the bedrock upon which the laws of traditional science are built. As such, they permeate to this day industrial societies' public knowledge about the environment. This means, environmental mat-ters are viewed through this lens and accorded legitimacy only through the mediating loop of science: politicians defer to it for definitions of safety and proof, medicine is steeped in it, technological development is the material-isation of its theories. The pervasiveness and power of that vision is detailed in the pages of this text. At this point I therefore merely want to set out the understanding of time that underpins Newtonian praxis. I have written about these matters in other contexts (Adam, 1988, 1990) but since the Newtonian heritage and its particular approach to time are so central to the environmental concerns of this book, it seems important to outline some of its key features at the outset.

## De-temporalised time: the Newtonian heritage

Scientific disembodied journeys to the frontiers of knowledge and the realm beyond time have their precursors not just in the development of the linear perspective but reach way back in time and across space to the traditions of magic and animism and the way of the shaman (see also Sheldrake, 1990 Chapter 2). While the flight from the limitations of body and earth is the very essence of shamanic power, scientists today are expected to reach into the deepest recesses of the invisible world and render it visible. They achieve this feat with the aid of technology, extending their territory to encompass the spatio-temporal world beyond the senses, providing disembodied knowledge of stars and atoms, ancient pasts and nuclear futures. 'But the ideal', as Sheldrake (1990:43) notes, 'is not confined to the ranks of professional scientists and technocrats; it has an all-pervasive influence on modern society.' Like a dye that has seeped through the fabric of industrial life, the scientific paragon of atemporal, detached, objective knowledge has become associated with the very meaning of 'truth' and legitimate knowledge. As such it accentuates the divisions between space and time and the power differentials between nature and culture, body and mind, everyday knowledge and science, green philosophy and environmental economics.

When time does enter scientific knowledge, it is in a form that bears little resemblance to the complex temporal times of nature and social organisation. Newtonian science recognises no contextually based differences in rhythm and intensity, no contextual tempo or timing, duration or change, no times inherent in processes and phenomena, no force that constitutes *natura naturans*, no *Wirkwelt* that works below the surface, no life and death, growth and decay, no seasonality, no right time for every season and place, no special days and moments or difference between sacred and profane times, no stress and pressure of 'deadlines', no decorum, no valorisation of speed, no reverence for the past, no hopes and fears for the future. Instead, time in Newtonian science is tied to the measure of motion: when something moves it covers distance which takes time. It is time taken, the measured duration between frozen events, the mathematical statement of acceleration and slowing down, of rates of change, of the difference between before-and-after measurements of fixed states. It is an atemporal time, a time unaffected by the transformations it describes. Moreover, the Newtonian physicists' equations are reversible, indifferent to unique spatio-temporal locations, to the unidirectionality of energy exchange, to growth, ageing and decay, to the cumulation of knowledge.

Even today, the sought-after ideal of Newtonian physics is independence from time and space, an atemporality which renders change static. The process by which this understanding is accomplished is one of extracting parts from their interactive, interconnected wholes, seeking the unit-part in its ultimate simplicity. That achieved, motion is not only measurable and predictable but reversible. That is to say, if one excludes friction, if one excludes gravity, if one excludes electro-magnetism, if one excludes interaction, if one excludes context and boundary conditions, if one excludes life, if one excludes knowledge, if one excludes any kind of human activity, emotion, interest and frailty, if one excludes all that, then one is left with a universe of perfect symmetry, single parts in predictable, uniform motion and reversibility. Bereft of direction, time is de-temporalised and the world subject to laws and predictability. In this physical world of single parts in motion everything is present now; change is merely the rearrangement of already existing bits. This means knowledge is a problem of quantity, patient accumulation of discoveries is the path to enlightenment and truth. From a

Newtonian perspective, therefore, uncertainty is a mere matter of temporary ignorance, a state to be overcome, as mentioned above, by further scientific research and technological development.

The ultimate expression of the Newtonian understanding of time is the clock where time is rendered visible as motion in space, tied to quantification through the number system, and abstracted from context. Clock time is thus invariable, standardised and universally applicable. In a more general sense, however, Newtonian time is invoked whenever technology is designed atemporally for specific functions only, that is, without concern for the life-cycle of a product, without reference to its interaction with the environment, without recognition that the created artefact forms an integral part of the world-wide web of interconnected processes. That is to say, technological products are premised on the Newtonian principles of decontextualisation, isolation, fragmentation, reversible motion, abstract time and space, predictability, and objectivity, on maxims that stand opposed to organic principles such as embedded contextuality, networked interconnectedness, irreversible change and contingency.

But, of course, irrespective of their design principles, technological products interact with their environments. This means, in their conception they may be decontextualised, isolatable and abstractable, but in their embedded functioning they interact and transact, leaving their irreversible mark. This disjuncture between conception and function which is such a central key to better understanding environmental pollution, hazards and degradation, tends to be left unattended in analyses. Instead, technology's negative impact and its discordant role in nature have led to calls for a 'cleaner', 'greener' and 'safer' technology, a technology that wreaks less havoc and works with rather than against the environment (see Weizsäcker, 1994). Without attention to the root problem of technology, however, it is difficult to think what green technology in both its reactive and proactive form could possibly mean. That is to say, technology will always remain a 'foreign body', to use a medical metaphor, as long as it is not conceived as integral to the ecology of life, as long as it does not encompass in its design the key characteristics of life, that is, temporality and contextuality, implication and immanence, the networked interconnectivity of give-and-take relations where the waste of one being and process constitutes the source of life for another (for a more detailed discussion on this subject see Adam and Kütting, 1995). The 'greenness' of technology based on Newtonian science assumptions is questioned throughout this book and the focus on time is used to bring to the surface its rationale, its future potential and its limits.

It matters how time is understood. It makes a significant difference, I want to insist, whether or not de-contextualised, abstract atemporality is aspired to, whether or not reversibility is a basic assumption, and whether or not the control of time and the transformation from transience to permanence are considered possible in principle and desirable in practice. To impose

41

an abstract, atemporal value on something transient and perishable, for example, is a cultural means to achieve permanence, to cheat death and the process of ageing, to pursue a form of alchemy. Similarly consequential is the assumption of reversibility. With respect to environmental hazards and degradation it entails the belief that mistakes can be undone, that increased knowledge and better technology can put right in the future mistakes of the past and damage inflicted on the environment now through pollution and the degradation of non-renewable resources. It presupposes that further research, science and technology will be able to undo the damage to the ozone layer, reverse the acidification of soil, water and plants, get rid of BSE and restore the British beef industry to its former glory. The assumption of reversibility implies, however, that it is possible in principle for ashes to turn back into burning logs and to re-attach themselves to the living trees they came from, for baked cakes to separate themselves back into their ingredients, for people to get younger and artefacts to get newer not older. Put like this, the belief in reversibility is clearly absurd, an impossibility outside the abstract realm of Newtonian theory, outside the conceptual scheme that depends on every connecting thread to the interactive reality of ecological interdependency to be severed. And yet, reversibility is a widely held assumption that permeates environmental discourse, surreptitiously facilitating risk-taking rather than precaution, and encouraging big science projects at the expense of ecological design.

Equally far-reaching is the assumption of objectivity with its dependence on the decontextualisation and externalisation of both 'observers' and the environmental matters in question. The transformation of persons from now—here participants to no-where observers and the transposing of now-here industrial hazards to external global 'problems' removes the urgency from environmental matters and allows them to be shunted along to become generalised 'problems' for others, environmental rather than existential matters for concern. In a context of global information, finance and environmental hazards the boundaries between societies, nations, institutions, disciplines and socio-economic—political activities are increasingly difficult to maintain. At that level, therefore, the concept of 'the other' loses its meaning. This applies most pertinently to pollution emanating from the industrial way of life. 'With nuclear and chemical contamination', Beck (1992b:109) thus suggests, 'we experience "the end of the other", the end of all our carefully cultivated opportunities for distancing ourselves and retreating behind this category'. With globalised hazards there is no position outside the frame of reference, no place from which to objectively observe. Everyone, every being, every thing, present and future, is implicated, if in exceedingly uneven positions of power. One would have expected therefore that members of industrial societies and their political leaders would recognise 'the soiling of the nest' syndrome, would realise that they are destroying not others' but their own bases of existence, and would consequently take the necessary steps to remedy that situation, that is, go for renewable energy production, take a precautionary attitude to new technological inventions, prioritise reproduction over production, shift towards eco-design with its goal of no waste and no pollution. This clearly is not the case. In the contemporary global context of implication, the linear-perspective and Newtonian habit of objectification and externalisation persist: objectivity continues to be a key precondition to legitimate knowledge; the externalisation of environmental degradation, pollution and other hazards is maintained as a taken-for-granted norm. Ameliorative action, as I show in this book, is conceived and constructed accordingly. Everywhere we look, Newtonian science assumptions reign supreme; they have become naturalised as an unquestioned habit of mind, foreclosing potential options and choices for trans/actions more appropriate to contemporary conditions.

It is far from my intention to suggest that there is a 'reality out there' that is falsified by the prism of this perspective. What I do want to argue, however, is that knowledge is always mediated, be this through visual representation, stories, metaphors or perspectives. This means, we are inescapably dependent for comprehension on a conceptual tool. We need mediators to turn life into 'reality'. If we now find that externality, decontextualisation and emphasis on vision and space are inappropriate conceptual tools for the temporal complexity we seek to understand – the *Merk-* and *Wirkwelt*, *natura naturata* and *natura naturans* – then we need to break through the shield of these taken-for-granted assumptions. We have to turn the reflective attitude on our own knowledge bases and begin to question the 'natural' place of these presuppositions in our ways of knowing, that is, our epistemology, ontology and cosmology. Recognition of the shortcomings of the tradition, however, is not enough if it is not simultaneously accompanied by a search for alternatives. In the quest for more appropriate bases of understanding and environmental action there is a tendency to seek inspiration from temporally and spatially distant sources, from ancient magic and pre-Enlightenment thought, from Eastern philosophy or the traditions of American Indians and Australian Aborigines. My own preference is to draw on resources available within industrial cultures here and now. Before I focus on some of the powerful knowledge bases of everyday life, however, I want to explore briefly the scope of post-Newtonian science for a potentially more meaningful engagement with the *Wirkwelt* of *natura naturans* and the environmental timescapes of modernity.

## Beyond Newton: engagement with multiple temporalities and the invisible

In Newtonian science there is no room for uniqueness and creativity, contingency and contextuality, surprises and discontinuities, chaos and catastrophes. Newtonian science has no means to encompass the invisible and

43

immaterial, the soul of nature and the 'supernatural', the temporality and immanence of *natura naturans*. Significant scientific developments since Newton have whittled away the extremes of this understanding of nature and brought scientific theory closer to everyday life and experience. Whilst they have not managed to incorporate temporality and im/materiality beyond sense data fully within their frames of reference, some of the major innovations in natural science praxis since Newton have engaged with different aspects of it: Evolutionary Theory through its emphasis on historical change; Thermodynamics through its stress on the irreversible arrow of time; Field Theory through its acknowledgement of invisible forces; Chaos Theory through its shift towards non-linearity, relativity and the importance of initial conditions; Quantum Theory through its account of fundamental temporality and virtual reality; and, finally, holographic technology through its reconfiguration of the relation between parts and wholes and its conceptualisation of enfolded implication. All engage to some extent with invisibility and immanence. All reconceptualise the temporality of material being.

Thus, Thermodynamics, for example, introduces irreversible directionality to the understanding of physical processes. In keeping with the Newtonian tradition, the First Law of Thermodynamics states that the total amount of matter and energy is constant, that it can be neither destroyed nor created but merely rearranged and conserved in different forms: nothing can disappear and conversely nothing new can be added. The Second Law, however, places an important restriction on this idea of endless transformations. In distinction to Newtonian dynamics it points out that energy is conserved but not reversible, that within this fixed total the transformations are unidirectional from order to disorder and waste, from cohesive energy for work to energy dissipated into the environment. The Second Law thus resonates with the temporal experience of unidirectionality – of ageing and decay, of logs burning to ashes, sugar dissolving in a cup of coffee, cars turning petrol into exhaust fumes, industrial processes producing pollution and waste – and with the knowledge that the reversal of these processes is an impossibility. In the mid-1980s, in an innovative study, Jeremy Rifkin and Ted Howard theorised the connection between the Second Law and pollution.

> Many people think that pollution is a by-product of production. In fact, pollution is the sum total of all available energy in the world that has been transformed into unavailable energy. Waste, then, is dissipated energy . . . pollution is just another name for entropy; that is, it represents a measure of unavailable energy present in a system.
>
> (Rifkin and Howard, 1985:45)

Despite its obvious relevance for understanding one of the key physical relations between industrial activity and pollution, however, knowledge of Thermodynamics is far less prevalent as a tradition of thought than its predecessor. Compared with Newtonian science, its permeation of everyday knowledge and thus its use in public environmental discourse is insignificant. (For a detailed account of Thermodynamics and its implications, see Adam, 1990:61–9; Briggs and Peat, 1985; Rifkin and Howard, 1985.)

Beyond the conceptual shift from reversibility to unidirectional irreversibility Thermodynamics changes the time focus from the external measure of motion to the internal temporality of processes and phenomena, from Newton's t-coordinate to the T-operator of historical change and the difference between past and future. Whilst the initial development of the theory covered cumulative, linear-progressive change, the Nobel laureate Ilya Prigogine turned his attention to non-linear, chaotic processes. To this end he developed the Theory of Dissipative Structures which covers living processes and flow structures (see Prigogine and Stengers, 1984 for an introduction to Prigogine's work). His is a science of becoming which needs to be differentiated from the science of machines: whilst the efficiency of a machine is judged by how close it is operating to equilibrium, living systems and flowing structures function far from equilibrium. It is at the extreme points, Prigogine argues, that we find the potential for creativity, that is, the capacity of the entire system to 'flip' into new levels of order. Newton's linear, time-reversible world is without surprises, a machine reality that has the potential to be known completely and to be taken apart and rebuilt again. In the living world of dissipative structures, in contrast, far-from-equilibrium processes are 'extravagant' in their energy exchange, dramatic in their change and unpredictable. They generate high levels of entropy which is not wasted but exchanged as a creative source of order in the give-and-take of life. Prigogine has legitimated the directional time of history for the physical sciences. He has established this irreversible, unidirectional, internal temporality as a law of nature and with it has changed the nature of a scientific law. Moreover, his work like that of chaos and quantum theorists incorporates the scientist's framework of observation in the analysis. No scientific activity, he insists, is without communication, time orientation, distinction between earlier and later states, past and future extension, historical embeddedness and contextual embodiment. The reality scientists seek to understand must therefore include themselves in the object of their study. I shall periodically return to these important issues in later chapters.

Prigogine's work represents one strand of what is popularly known as Chaos Theory. Another branch relates to the mathematics of fractals, of structures whose unpredictable complexity is constituted not through the dramatic energy exchange of dissipative structures but through vast quantities of interdependent and reflexively organised repetitions. In this mathematical approach the visual–spatial emphasis of the linear perspective and Newtonian science is retained through the computer representation of the exceedingly complex mathematical models, as is much else of the deeper structure of Newtonian assumptions. Some of the scope of the innovative theoretical insights of chaos theory for understanding environmental processes is discussed in Chapter 6 with reference to radiation. At this point I merely want to identify through the example of hormone-disrupting chemicals a number of temporal features such as non-linearity, the importance of initial conditions and non-proportionality which depart from the Newtonian and linear-perspective vision. (For my examples I draw on the seminal work of Colborn *et al.*, 1996 while my adaptation of chaos theory for understanding socio-environmental processes is based on years of engagement with the subject. Excellent popular accounts of chaos theory are provided by Briggs and Peat, 1985 and Gleick, 1987.) In order to avoid tiresome repetition when I refer to scientific perspectives beyond Newton in the chapters that follow, I draw an initial sketch of their contours in this introductory conceptual part of the book.

In many ways, chaos theory inverts Newtonian logic. Instead of abstraction and linear cause-and-effect chains this theory focuses on interdependent, contingent connectivity. Instead of certainty and predictability it provides a rationale for the impossibility of establishing connections backwards from symptoms to cause or forward from action to future outcome. Where Newtonian science assumes single parts in motion, chaos theory presupposes complexity and associates this with continuous, iterative interaction (that is, where the repetition of actions involves feedback). A chaos perspective recognises that there are innumerable paths and outcomes from the same initial conditions and that the same symptoms can have radically different originating factors and paths leading to it. It appreciates that change in one part of the system reverberates through its entirety – as minor ripples and/or massively amplified – in some instances facilitating major change, in others barely registering a difference. In living social systems where interactions occur in a time-layered fashion at every level of existence, from cells to stars, identical paths and outcomes would constitute the equivalent of a miracle. From this perspective one does not assume linearity, just as one does not expect to establish causal connections backwards from a hurricane to a specific butterfly flapping its wings. Instead, one is *amazed* to find clusters of look-alike systems where interactions have seemingly progressed along the 'same' paths from likely similar initial factors and conditions. Based on an assumption of complexity and non-proportionality, the quest is for patterns and correlations. From the vantage point of chaos theory, therefore, the socio-political quest in environmental matters for proof and certainty is meaningless, a futile exercise in public relations that is bound to fail.

Recent research on hormone-disrupting chemicals (Colborn *et al.*, 1996) can serve as an exemplar for environmental hazards that are beyond the reach of Newtonian science explanation and for which chaos theory provides in parts a more appropriate conceptual base. This body of research which has been gathered and collated from isolated and globally dispersed sources indicates on the one hand an increasingly impaired reproductive capacity across species and on the other a world-wide incidence of synthetic chemicals stored in body fat. The bringing together of this diverse body of research showed for the first time a) that symptoms associated with the disruption of the endocrine system are similar across species and b) that the resulting damages originate with exposure in the womb to synthetic chemicals that mimic natural oestrogen. It appears that no place on this earth is safe from contamination: hormone-disrupting chemicals are found in animals living on the highest mountains and in the depth of the oceans, in members of industrial societies and natives of equatorial rain forests and the polar regions. Moreover, symptoms associated with impaired functioning of the endocrine system encountered in the young adults, it seems, have been set in motion during critical stages of foetal development. The connection between impaired reproductive capacity, synthetic chemicals stored in body fat and exposure to a range of chemicals at critical phases of early development has been established without any possibility of tracing back with certainty the path(s) from symptoms to specific source, from individual reproductive failure to a single chemical compound. That is to say, scientists have been able to identify the source of this pervasive health hazard but have no means by which they could provide 'scientific proof' of the connection.

> [W]here humans and animals are exposed to contamination by dozens of chemicals that may be working jointly or sometimes in opposition to each other and where timing may be as important as dose, neat cause-and-effect links will remain elusive.
>
> (Colborn *et al.*, 1996:196)

This in turn has socio-legal consequences since, in circumstances of this kind, the polluter-pays principle cannot be evoked. With hormone-disrupting chemicals (as with radiation, BSE and genetically modified organisms, the other examples of invisible hazards focused on in this book) societies the world over face a new kind of hazard. The pollution associated with hormone-disrupting chemicals has to be distinguished not only from earlier chemical poisons that killed outright when they were accidentally ingested but also from drugs such as Thalidomide that maimed and caused deformities to limbs and organs. In contrast to these earlier chemical threats, the dangers posed by hormone-disrupting chemicals are constituted in a far less direct way. They are marked by latency and remain invisible for long periods of time. Moreover, the dose is not proportional to the symptom and no single chemical can be tied to specific consequences. Finally, a variety of disabilities can arise from the same chemical contamination and similar symptoms can have numerous histories ranging from single and combinations of synthetic chemicals to unspecified other endo- and exogenic factors.

When in 1962 children were born with limb and organ deformities it was possible for scientists to establish the connection not just between the birth

47

defects and the drug Thalidomide but also between the time the mother took the drug and the period during pregnancy at which the drug did its maximum damage. Moreover, the afflicted group was large enough and the medical profession networked sufficiently so that a pattern could emerge and action could be taken accordingly. When these conditions no longer hold, we (scientists, politicians, and the public at large) are facing a qualitatively new problem that needs to be approached with assumptions other than linear causality and proof, objectivity and independence from context, simplicity and single parts in motion, reversibility and quantification. When the time–space link is broken and when the gap between the *Merk*- and *Wirkwelt*, between perceivable symptom and precipitative action-force, expands over a lifetime and is distributed across the globe, then recourse to the conceptual schemes of the mechanical age are more than useless – they are dangerous: they make us ask the wrong questions and look for futile solutions, and lull us into destructive complacency.

In 1938 British scientists developed a synthetic chemical that acted like natural oestrogen. This was hailed as a wonder drug that was to cure all ills and problems associated with reproduction, ranging from the prevention of miscarriages to birth control 'the morning after'. This chemical, known as diethylstilbetrol or DES was, however, far from safe and was doing its damage invisibly over a period of fifteen years and more before it materialised into a variety of symptoms during early adulthood. With such a long latency period, the link between poor health and a wide range of reproductive problems was no longer connectable to a pre-natal event in the way it was possible in the case of the Thalidomide drug. As Colborn *et al.* explain,

> [I]t is possible no one would have ever figured out that DES was doing profound but invisible damage to those exposed in the womb. Until DES, most scientists thought a drug was safe unless it caused immediate and obvious malformations. They found it hard to believe that something could have a serious long-term impact without causing any outwardly visible birth defects.
>
> And even when one recognises that prenatal events can lead to medical problems years later, a long time lag between cause and effect makes it difficult to prove connections or even verify that the mother had been exposed to the suspected drug or substance. In the case of DES, fears of liability on the part of doctors have only added to the difficulties of those exposed to the synthetic estrogen.
>
> (Colborn *et al.*, 1996:53–4)

Moreover, with DES there are a number of further factors that take this health hazard outside the bounds of conventional science assumptions. First, there is no 1:1 relation between drug and symptom: the sons and daughters of women who took DES during early pregnancy suffer from a variety of afflictions and debilitating symptoms which range from disabilities associated with the immune system and chronic allergies to damage to the reproductive system, including a variety of tumours and cancers. As I indicated above, this is a typical characteristic of chaotic systems, where one action can result in numerous outcomes and the same symptom can have a variety of different factors and paths giving rise to it. It seems doctors got alerted when an extremely rare cancer which had previously occurred only in women over fifty years of age suddenly afflicted a number of young women aged between fifteen and twenty-two years. It was the rarity of this event that set in motion the search for connections that eventually led to DES. However, since not one but a variety of symptom clusters are associated with this drug, the exclusive causal link to DES could never be confirmed. Consequently, the question remains officially unresolved. A second feature relates to the dose-symptom relationship and to the centrality of timing. The severity of the symptoms is not related to the dose: there is no 1:1 relation between dose/quantity and impact. Contrary to traditional quantitative perspectives on chemical pollution and poisons, a large dose does not lead to an equivalent big effect. Instead, extremely low doses can have devastating life-long consequences if the drug was taken by the pregnant mother during critical stages of her foetus's development. Thus, if the quantitative approach is retained in scientific safety trials and if the research does not discriminate between early, middle and late phases of pregnancy then the magnitude of the drug's effect on the unborn is masked. The findings are effectively falsified. Finally, the DES case brings to the fore the interconnectedness of the body's systems. Colborn *et al.* (1996: 62) report on animal studies which indicate that disruption in one part has unpredictable effects throughout the body over an organism's lifetime.

> DES acts on other parts of the developing foetus besides the reproductive tract, including the brain, the pituitary gland, the mammary glands in the breast, and the immune system, causing permanent changes there as well. Researchers have found evidence that prenatal and neonatal exposure to DES or other estrogens can sensitize the developing foetus to estrogens and perhaps make it more vulnerable later in life to certain cancers.
>
> (Colborn *et al.*, 1996:62)

Since, however, these processes are time-distantiated, that is, marked by long time lags, the language cannot be one of certainty, truth and proof but is necessarily permeated by uncertainty, indeterminacy and consequently caution and suggestiveness. It could not be otherwise since proof and certainty of the Newtonian kind are unavailable in principle in time-sensitive, time-distantiated environmental hazards of this kind.

The level of complexity and indeterminacy is moved up a further notch when we examine the effect not just of one drug on one species but of a group of drugs that act globally across the breadth of species, when the chemicals-to-symptoms link is open with respect to time and space and when patterns therefore have to be built up for an indefinite future across national and disciplinary boundaries. Quite apart from the difficulty of finding suitable conceptual frameworks within which to study such complexity, the vastness of the time–space distantiation in conjunction with the traditional organisation of knowledge into separate disciplines impede the building up of appropriate knowledge for such pressing environmental issues. Of course, as I shall argue and indicate throughout this book, questions of economic, political and scientific interest and power are never far from the surface. At this point, however, I am concerned to elaborate some of the time–space complexities of largely invisible environmental hazards and the substantial difficulties these pose for conceptualisation, theory and research methodology. In the case of hormone-disrupting chemicals this involves: life-long impacts and time–space distantiation; vast time-lags and periods of invisibility; critical periods, amplification and contingency; time-layered and delayed symptoms; non-linear change and progression paths; trans-generational damage and non-proportionality between chemicals and symptoms, dose and impact; no control group with which a 'healthy norm' could be measured; interdependence and networked connectivity of chemical processes, body systems and environments; and, finally, an inescapable multiplicity of mutually influencing chemicals and the consequent impossibility of functionally isolating single compounds.

Where Newtonian assumptions and the linear perspective fail, Thermodynamics, the Theory of Dissipative Structures and Chaos Theory offer some important insights which contribute to the understanding of the bewildering complexity. Singly and collectively they provide points of departure and move towards recognising multiple temporalities, irreversibility, indeterminacy, and the centrality of the invisible. Of equal importance, however, is the need to engage with the transcendental realm, the encoded 'reality' beyond the reach of the senses which is constituted by both a potential and invisible *Wirkwelt*, an enfolded future and *natura naturans*. Here, the post-Newtonian scientific perspectives have not that much to offer; only creative adaptation and substantial extension of the theories can guide understanding in that particular direction. (My focus is on theories of the physical sciences not because I think them to be superior to other theories but because they are accorded the highest prestige and status at all levels of sociopolitical action, media representation and in everyday life – because they are equated with truth and the description of reality.)

While there may not be a suitable theory in the contemporary physical sciences, there is a technology – holography – that provides a powerful metaphor not only for encoded 'reality' and enfolded potential but also for the mutually implicating relationship between the part and the whole.

Technologies, it needs to be appreciated, are not just 'dead things' whose sole function is to be used. Rather, throughout the ages, such artefactual creations have served as a means for understanding ourselves and the environment. The clock, the steam engine, the telegraph system, the computer – they all have been used as bases for understanding. From the clockwork universe to the computer brain, technologies have influenced how we see and understand the world. They have become metaphors, habits of mind that are inextricably linked to the way we relate to each other and the world around us. I have written about this technology–metaphor–practice link at length elsewhere (Adam, 1990, 1992, 1996b), so I do not want to replicate these arguments. Here, I merely want to outline some of the key points relating to the hologram and identify the relevance of this metaphor for understanding *natura naturans* and for approaching the complexity of the sort of environmental hazards I discuss in this book. Holography is a technology that encapsulates a post-Newtonian conception of the part to the whole and entails a process of representing and fixing reality visually that differs significantly from lens photography. Holography implicates the absent and centrally incorporates the idea of resonance. As such it reaches parts of the complexity that the above-mentioned theories cannot reach. To demonstrate the pertinence of this technology for understanding I need to plot it against the backcloth of lens technology and the latter's Newtonian conceptions and assumptions.

In lens photography the object stands in a 1:1 relation to the produced image and each point on the object corresponds to the same single point on the representation. In case of the plate being broken, the broken off part would be missing from the image. It means, this design principle of lens technology is underpinned by classical Newtonian science assumptions about the relation of part to whole, object of investigation to reality and, by implication, individual to society, being to environment. It links to the understanding that when something is removed – from a photographic plate, society, nature – the rest remains undisturbed; it is simply missing. Moreover, the part that is removed is assumed to remain intact, to retain its integrity and full functional capacity. Animals or bits of organism studied in the laboratory, single substance pollution studies, the quest for proof based on single, mono-causal connections all operate with this particular part– whole conception. The Theory of Dissipative Structures, and both systems and chaos theory have moved away from these assumptions and understand the part–whole relation as one of infinite connections where disturbance in one part affects the whole in non-linear, unpredictable ways. Holography functions on the basis of yet another conception of the relation between part and whole. As a technological metaphor it revolutionises this particular understanding and, like the post-Newtonian theories discussed above, it brings scientific–technological understanding closer to everyday experience and commonsense knowledge.

In holography the emphasis is on parts being implicated in the whole and

51

vice versa, on interconnectivity and on meaning arising from interaction, on multiple perspectivity and on the centrality of that which is not visible – what a gem of a metaphor for conceptualising environmental hazards! Derived from the Greek 'holo' which means whole and 'gram' which means to write, a hologram 'writes the whole' (Bohm, 1983:145). It implicates the totality in every tiniest aspect of itself by encoding its information not in the parts but in the pattern that arises from the interference created by two light beams. The light beams which are fired in phase get split: the one gets sent directly to the plate, the other encircles the object and picks up the reflections from its entirety before it is superimposed on the reference beam. Once the beams are reunited they are no longer in phase but interfere with each other. The multi-perspective vision which is plotted against the 'non-involved' reference view results in a pattern that bears no resemblance to the object but has its features encoded on the basis of difference. A three-dimensional image is then created from the encoded pattern by shining the reference beam on the plate and viewing it from the other side. This means, the 'memory' of the entirety is recorded throughout the plate; each part of the photographic plate thus contains the whole. If a piece is broken off from a holographic plate, therefore, the whole is preserved in both the broken-off piece and the rest of the plate whilst clarity and definition are diminished. This means, in holography 'parts' cannot be extracted or abstracted as parts from the whole. The whole is always both implicated and affected.

This technology is so rich in metaphorical implications that it is difficult to know what to mention first: the importance of interaction, the multi-perspective vision, the encoded meaning, the inviolability of the whole, all vie for pride of place. Where lens technology emphasises the part, holography images the whole, it reconstitutes an indivisible, inviolable whole. There is nothing else. The holographic whole is reconstructed through a reference perspective which combines with a multi-perspective vision. This process involves difference and comparison, stability and change, visibility and invisibility, encoding and decoding. Out of the relation between observer and reference perspective arises the representation of reality. This means holography fundamentally implicates the perspective of the observer. There is no choice here between objectivity and relativity, neutrality and bias: if all processes occurred in phase we could not know them in relation to each other. We would have no basis from which to recognise difference and change. Finally, even at the speed of light, what matters with reference to the constitution of this image-reality is the invisible bit between action and product, the *Wirkwelt* operating between input and output. Focus on that *Wirkwelt* highlights the pivotal role of interaction, of the 'observer' perspective, of interference and disjuncture, and it brings to the fore a whole set of new fundamental concepts such as implication, resonance, contingence, enfolding and encoding. Their centrality for understanding environmental processes is demonstrated throughout this treatise.

If we now look back on these theoretical and technological moves away from Newtonian science and the linear perspective we can see that each of these scientific innovations harbours important insights for environmental analyses. Singly they may not amount to much but considered together they offer an impressive spectrum of conceptual change. Collectively, they facilitate a shift in understanding towards appreciating the importance of the in/visible, im/material, latent, encoded, immanent 'reality' that receives such short shrift in the materialist perspective. Taken together, they reduce the culturally constructed distance between us and the hazards, and show our implication as perpetrators and recipients. Finally, they put science on a new footing *vis-à-vis* the environment. This is of crucial importance in a socio-political context where natural science (which tends automatically to mean Newtonian science, irrespective of the scientists' own perspective) is equated with legitimate knowledge. In the UK, for example, recourse to science in times of trouble, such as the BSE crisis (see Chapter 5), has been taken to a point where politics and policy making are abdicated to that branch of knowledge production. With environmental hazards increasing and intensifying year by year, it is essential to strengthen post-Newtonian scientific theories and to legitimate associated new base assumptions. Vindication of these new presuppositions, however, must not result in their blind acceptance and naturalisation. They must not become new habits of mind. Instead, they need to be used consciously and reflectively as conceptual tools, be chosen selectively for their appropriateness to the tasks in hand. Their suitability, therefore, needs to be checked against everyday experience and commonsense knowledge. Even though this experiential knowledge is permeated by the metaphors and assumptions that are materialised in the artefacts we interact and live with on a daily basis, this knowledge also transcends them: it is the source of creativity and novelty, the wellspring from which innovations arise and the root of wisdom that needs strengthening in the effort to shift from material, spatial ways of knowing to a deeply temporal and transcendental perspective more suited to engage with the *Wirkwelt* of *natura naturans*. Since everyday knowledge makes for a potent antidote to the seductive power of the materialist world of visual certainty and control, I draw on it extensively to counter the hegemony of conceptual conventions and to strengthen alternative ways of understanding. Let me begin this process by elaborating one of the ways we deal with the problem of latency and invisibility on an everyday level.

## Environmental timescapes in a context of industrial culture

When we look at a landscape we see historical records of activity: of wind, weather and climate, of the growth cycles of nature, of animal and human life. Consistently lopsided trees, for example, indicate coastal winds and the nearby sea. Hedges and stone walls tell us about human agricultural activity even when there are no houses or people engaged in such activities to be seen. Moreover, by looking more closely at the stone walls, we can tell about the geology of the area, about the kind of farming that is practised in the locality, and about the animals that are being kept within those boundaries. The way the stone walls are built and maintained gives us a further indication about current activities and the state of individual farms.

A landscape therefore is a record of reality-generating activity. It is a chronicle of life and dwelling. (For innovative writings on knowing landscape, see Ingold, 1993; Shama, 1995.) That is to say, the visible phenomena making up the landscapes have the invisible constitutive activities inescapably embedded within them. The landscape thus includes in its representation spatial and temporal absences, the intrinsic *Wirkwelt*. It tells a story of immanent forces, of interdependent, contingent interactions that have given rise to its existence. From the point of view of the observer, of course, a landscape can never be an objective absolute, since what observers can see depends on their prior knowledge, their power of deduction and their imagination. The scape − be this a landscape, seascape, or cityscape − arises from the interactive unity of observer and observed, of material phenomena and forces inaccessible to the senses, of visible and invisible influences.

The following points are of importance here: a landscape is a record of constitutive activity. It includes absences. It combines natural and cultural activities into a unified whole. It is relative to the eye of the beholder. As such, everyday understandings of the landscape differ from images of nature and culture, discussed at the beginning of this chapter, which are defined negatively in relation to each other: culture as the product of humans and not nature, nature as that which is created without the aid of humans and which functions irrespective of or despite human activity. A landscape perspective, therefore, is inclusive; it gathers up sources of knowledge from both material and 'immaterial', visible and invisible sources. Such transcendence of materialist and dualistic approaches becomes crucial for understanding a world of globalised local human activity that creates holes in the ozone layer, changes the level of $CO_2$ in the atmosphere, damages the unborn and causes abnormalities at the level of cells in plants, invertebrates and humans alike. It becomes pertinent for a context in which nature is inescapably contaminated by human activity and this acculturated nature returns, boomerang fashion, as hazards, constituting the always pressing conditions for re/action.

It seems to me, the kind of knowing that is entailed in such a landscape perspective constitutes an excellent base upon which to develop a sensitivity to the complex temporalities of contemporary existence. With the idea of the timescape, I seek to achieve an extension of the landscape perspective, that is, to develop an analogous receptiveness to temporal interdependencies and absences, and to grasp environmental phenomena as complex temporal, contextually specific wholes. This involves a shift in emphasis not just from space to time but, more importantly, to that which is invisible and outside the capacity of our senses. It entails an engagement with the immanent *Wirkwelt*, which is so central to the understanding, conceptualisation and analysis of contemporary hazards such as hormone-disrupting chemicals, radiation and genetically modified organisms, and is such a crucial precondition to appropriate responses and actions. Since we have no sense organ for time, we need – even more than for the landscape perspective – the entire complement of our senses working in unison with our imagination before we can experience its workings in our bodies and the environment. Such an effort at the level of imagination is needed if we are to be able to take account in our dealings with the environment of latency and immanence, pace and intensity, contingency and context dependence, time-distantiation and intergenerational impacts, rhythmicity and time-scales of change, timing and tempo, transience and transcendence, irreversibility and indeterminacy, the interrelation between *Merk-* and *Wirkwelt*, the influence of the past and the projection into an open future.

A timescape perspective enables us to integrate scientific and everyday knowledge, and the constitutive cultural Self with the workings of nature. It facilitates a recognition of the clashes and stresses that tend to be left implicit in both classical science analyses and political debate. It allows us to move from single and dualistic approaches and abstract, functional perspectives to knowledge that emphasises inclusiveness, connectivity, and implication. It promotes understanding that acknowledges the relativity of position and framework of observation whilst stressing our inescapable implication in the subject matter and acknowledging personal and collective responsibility. It explicitly incorporates absences, latencies and immanent forces, thus helping us to move away from the futile insistence on proof and certainty for situations characterised by indeterminacy, time-lags of unspecifiable durations and open dispersal in time and space.

If we take once more the example of hormone-disrupting chemicals, a timescape perspective brings into a coherent socio-natural whole all the diverse and disparate time aspects of this particular existential hazard. It establishes connections between ineluctable multiple time-lags, the importance of timing and the persistence of the chemicals and it locates these processes within an intergenerational time-frame of analysis. It differentiates the hazards arising from these chemicals from other chemical poisons where shorter time-lags and less time–space distantiation allow for limited causal analysis and thus the potential at least for proof and recourse to the polluter-pays principle. It encompasses inescapable irreversibility and fundamental indeterminacy in conjunction with socio-political and scientific responses based on futile assumptions of potential reversibility and certainty, atemporal scientific truth and therefore possible proof. By emphasising the time distantiation not just into the future but also into the past a timescape perspective can explain why scientists and politicians are continuously called short, how it is that they can insist on absolute safety 'in the normal sense of the word' – be this about the use of chemicals, atomic energy, agricultural practices or the release of genetically modified organisms – and then, within days of such pronouncements, have to do a 180 degree turnabout and declare the 'scientifically guaranteed safe situation' a potentially devastating hazard.

Traditional scientific knowledge and environmental safety regulations tend always to be past oriented. They are based on the previous poisons and hazards, on that which is by then known and understood. It is within the nature of a *new hazard*, however, that it contains features and characteristics that are not known and differ from previous threats. By the time such a new hazard is understood, therefore, it is no longer suitable as a basis for preventing the next generation of consequences of technological innovation and the industrial way of life. Thus, in the case of hormone-disrupting chemicals, the test on adult males for signs of poisons that kill or maim was the application of inappropriate knowledge about past poisons to a hazard that is transmitted in the womb, a danger based on contact with exceedingly small doses at critical times during pregnancy which causes lifelong damage and disability that does not materialise as symptoms until early adulthood. With respect to such new hazards, socio-scientific response and political regulation tend to be fighting the previous war (Colborn *et al.*, 1996:242). On the basis of this past-oriented system of response, prevention and environmental protection, we can be certain that 'surprises' will be produced at ever shorter intervals and ever greater intensity. Moreover, as successes turn into hazards, each new threat seems to turn up at a time and in a place and form that was not expected. As Colborn *et al.* (1996:242) point out, 'the surprise will be something never even considered. If anything is certain, it is that we will be blindsided again.'

As such, a timescape perspective conceives of the conflictual interpenetration of industrial and natural temporalities as an interactive and mutually constituting whole and stresses the fact that each in/action counts and is non-retractable. This in turn has the potential to encourage more cautious, precautionary and sustainable re/action than is the case with an assumption of potential certainty and reversibility and it promotes recognition that our relationship to time is centrally implicated not only in the industrial way of life but also in any conscious construction of a sustainable future. Humans have always had an impact on nature, just as the body and the environment influence cultural development. The difference to earlier historical periods is one of scale, degree and reach of culturally constituted environmental hazards. That is to say, the impact of the industrial way of life is global in its reach and of a temporal scale previously unknown; as such it narrows choices and affects life chances of future beings from the next generation and from thousands of years hence. This contemporary projection into the future relates to the last Newtonian science-based habit of mind that I want to introduce in this first chapter.

## Dreams of safety, certainty and control

In the not-too-distant past, the future belonged to the gods. Today it is considered a human resource to be harnessed and exploited. As a challenge to human control, the contemporary future is predicted, calculated, insured, and discounted. Through industrial activities today, the futures of countless generations are predetermined, their options foreclosed for an untold number of years hence. The future is thus dealt with and eliminated in the present. This is not to suggest that the reach into the future *per se* is a phenomenon of the industrial age. Far from it. One could even say that the conscious extension into the future is a mark of culture, an essential aspect of what it is to be human. That is to say, the storage and preservation of food, the externalisation of knowledge in art, and the performance of rituals, are all evidence of the human transcendence of the present, dating back to prehistoric times. Through sacrifices and the creation of temples, humans attempted to influence the whims of the guardians of the future, of the gods who had sole rights to this dimension of people's lives. Through ritual and burial of the dead, they sought to pacify ancestors, affect potential actions from the otherworld. Through art and writing, culture could survive individuals into subsequent generations.

And yet, I want to suggest, there are significant differences between these above-mentioned means to transcend the present and industrial societies' approaches to the near and distant unknown. The latter's future belongs to the present. Safety, certainty and mastery are considered legitimate socio-economic and political goals which arose with both the increasing power of science to move back the frontiers of the unknown, innovative technology, and the development of insurance to provide financial compensation for spoilt futures. These particular ways of reaching into the immediate, near and distant future are based not on bargaining with ancestors and gods but on a techno-economic relationship to a resource that is to be used, predicted, allocated, managed, sold, colonised and controlled in the present. Thus, for example, the development of electricity was central for extension into the immediate future of daily existence. With its universal adoption by industrial societies, the night could be colonised: no longer the realm of darkness, mystery and fear, it became instead an extension to the time of work and pleasure (see Melbin, 1987). With electronic communication, distance lost its link with bodies moving across space. The near future was brought into the present. Once more, the colonial principle has been extended from space to time. While the compression of time, afforded by machine power, eliminated both immediate and near futures, other technologies created and subjugated the long-term, open future, pre-empting presents of countless successor generations of humans and other species. I am referring here once more to the impacts of synthetic chemicals, nuclear power and geno-technology, all technologies that express a belief in the potential knowability and controllability of the future: science may not be there quite yet, but, so this particular belief system insists, there is basically no cause for concern. The holy grail, the unifying theory, the solutions to the million-and-one hazards on the technological rebound, all are just around the next corner. Enlightenment and control are merely a question of time. The next round of technological fixes will sort out all current problems. For a socio-scientific perspective based on timeless laws, abstract particles in motion, reversibility and endless transformation from one state into another, the future is a realm of mere temporary uncertainty, open in principle to exploitation and control.

The technologies that have arisen from that science, however, have burst out of the limiting framework of their inception. They now confront their creators and those who champion their development with unprecedented conceptual dissonance: the future created through those technologies and on the basis of the assumptions discussed in this chapter is not a future managed and controlled, not the future of insurance and computer simulation, not the future of prediction and certitude. It is the future of manufactured risk and uncertainty, of time bombs threatening to go off. It is a world of blank cheques that have to be honoured by cultures at the receiving end of this 'development' and by their already implicated successor generations. The parasitic relation of industrial societies to future presents and the present futures of spatially and temporally distant others poses moral as well as political dilemmas. These I want to address in this book. The focus on time aids me in this endeavour. In the chapters that follow, therefore, I consider how well the current guardians of the future are managing their task. I scrutinise our future-making institutions – markets, politics, agricultural and food science, and the news media – and ask: 'How safe is the future in their hands?'

## Time for recap and reflection

The key issues discussed in this chapter relate to habits of mind that are central to the industrial way of life. These taken-for-granted assumptions, so my argument goes, are pivotal to the way we approach nature, technology and environmental hazards. Their deconstruction and the development of more appropriate conceptual tools is the task that has been started here and will continue to be a guiding thread through the chapters. I have identified with those industrial habits of mind an approach to nature as culture's 'other' and the environment as external. I have associated with them emphases on space, on the visual and the material, on linear cause and effect chains, on abstraction and single parts in motion, on externalisation and objectivity, on *natura naturata* and the *Merkwelt*. Equally important to these thought traditions, as we have seen, are the allied exclusions: time and multiple temporalities, context, all that is invisible and 'immaterial', connectivity and interdependence, *natura naturans* and the *Wirkwelt*. Intimately connected to this theoretical, conceptual framework, finally, is a particular relation to the future: expectations of safety, certainty, predictability, control and enlightenment. If this is not being achieved, the belief is that it will be in the very near future.

These habits of mind, however, are no longer adequate and appropriate for understanding and responding to the kind of below-the-surface, beyond-the-present, time-distantiated hazards that have arisen from this approach to nature and its associated way of life. When those Newtonian thought traditions are evoked in the face of such hazards, the disjuncture and discrepancy become apparent: scientific pronouncements on and political responses to dangers from radiation to BSE lose all credibility and turn the statements of those entrusted with public safety and well-being into a farce. At the same time, however, those inappropriate assumptions and expectations affect people's lives and livelihoods, their own and future generations' safety and posterity. It is therefore essential that we confront the disconcerting losses associated with a departure from the comforting world of the linear perspective and Newtonian science and come to terms with the discomforting and discomfiting alternatives. This entails that we let go of our reliance on certainty, externality, abstracted simplicity and control, and that we dismantle the boundaries that separate us from nature, other cultures and other species. We can then begin the process of reconceptualisation, engaging consciously and explicitly with the multiple temporalities of contemporary existence and the constitutive world beyond the senses without, that is, losing sight of the centrality of interests and power. The exhilaration of flying blind may then give way to the excitement of a cautious re/construction of sustainable futures.

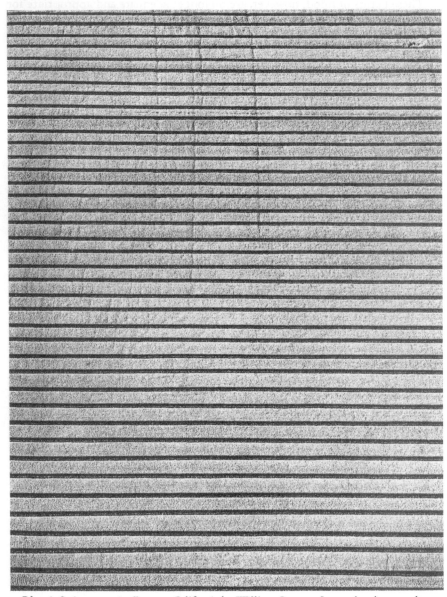

*Plate 2* Strip cut grain, Esparto, California by William Garnett. *Source*: the photographer.

## 2

# IT'S ALL ABOUT MONEY, ISN'T IT?

### Time, all things green and profitable, and moonlighting for the environment

## Introduction

Today, nature is big business: being green is sound economics. Green technology is a major development area. Environmental services are booming. Waste management is one of the largest growth areas. The 'green' revolution in high yield crops has transformed Third World agriculture. Environmental research is buoyant. Business, we can see, is taking the environment seriously. If the above examples are not sufficient, the new terms that combine the environment with markets unequivocally announce the unlikely marriage: the 'green market', 'green globalism', and the 'global environmental market' are just some of the conjunctures that indicate that money and economic considerations are inescapably tied to all matters environmental. What has happened then between this incorporation of the environment into the economic fold and the exploitation of nature as a resource, the recognition of damage to the environment, and the rejection of sustainable technology on the grounds that it is too expensive? Andrew Ross describes this shift in the following terms:

> Once the source simply of raw materials to be extracted and transformed into commodities, nature is now a valuable origin of exchange value in its own right. In the age of environmental accounting, nature enters the market not just as source of property or mineral value, nor just for its capacity to sustain its soil, water, and air, but also for its own sake, as a desirable signifier with inherent worth and value to consumer markets.
>
> (Ross, 1994:8)

Ever since the Brundtland Report (World Commission on Enviroment and Development, 1987) the marriage between economy and ecology has been sanctified, their interdependence officially acknowledged. How that marriage is conceived and consummated, however, makes a crucial difference to the way we respond to environmental degradation and the production of hazards.

I am not an economist, so a special amount of reading and thinking had to be done before I could write about the temporal issues entailed in this topic. As with all the other chapters, my discussion is primarily an attempt to bring to the fore the complex temporalities entailed therein and through that focus to shed new light on the issues. It is not therefore a critical consideration of environmental economics *per se* but an exploration of the assumptions and habits of mind that guide actions within and outside the time-economy of socio-environmental relations. As one would expect, money plays a central role in those relations. In the first part of this chapter I therefore establish a link between 'nature = money' and 'time = money' and locate both these assumptions in very ancient cultural desires, desires that are particularly well served by the mutually supporting conceptual habits of neo/

classical economics and Newtonian science. Next I consider the limitations of that vision and what might be involved if technologies were designed and businesses conducted according to ecological principles, that is, deserving of the right to call themselves 'green'. In the last part of the chapter I show that such ecological practices are pervasive at the level of daily life, that is, in all spheres where money does not enter as a medium of exchange. It is to everyday knowledge and practices therefore that we need to look for inspiration at times when sustainability threatens to become an ever less attainable goal.

## Nature = money, time = money

### *Nature = money*

When nature is equated with money, it can mean a variety of things. It can signify that nature is a resource which can be turned into money, that the study of nature can be lucrative, that money is to be made from being green in practice and product, that the protection of nature is financially attractive, or that all things natural have become valuable commodities. The temporal issues involved here are not easily recognised until, that is, we put the nature = money alongside the time = money equation. Placed next to each other we can see parallel developments and a shared pool of encoded desires, all with implications for attempts at changing current conceptual habits and economic practices.

Throughout history, humans have lived in, of, with and as an integral part of nature. Equally, cultural development entailed a certain degree of distancing from nature by entering in a relationship with nature, thus precluding the 'natural' way of life and unmediated knowledge. The gap between this kind of inevitable cultural distance from nature and the scientific approach to it, however, is enormous. It is the difference between being reflexively and contextually embedded to standing outside and conceiving of nature initially as a realm for observation and discovery and later as a resource to be harnessed. From there, the step to commodification is a small one. That is to say, the change from seeing nature as a resource to be harnessed to viewing it as a resource to be exploited for the accumulation of wealth is achieved with ease. The equation of nature = money becomes naturalised. When nature is conceived economically as an exploitable resource, that is, when it is seen in relation to its money value, then it follows that any unused nature means a loss of revenue, a missed opportunity for financial growth, money wasted.

With an example about fishing, Hans Christoph Binswanger (1994:178–82) shows the transition from living with and of nature to nature's economic exploitation, to nature = money. In a small community organised on the basis of subsistence economics, the family in his description used to fish for their personal need only. To take from the lake more fish than would have been absolutely necessary to feed themselves would have been considered not only greedy but also a sin against nature. The son, however, had secretly begun to fish more than the family needed and was taking his surplus to the neighbouring town to sell. The ability to exchange something perishable against money, that is, against a value that does not rot and can be accumulated and exchanged against a number of other goods which would not normally be accessible to this small community, makes the fishing for surplus an attractive proposition. Despite its appeal, however, there is not yet a compulsive element to this practice. That is to say, as long as fish is merely exchanged via money for goods of equal value, there is no pressing need to continue fishing for surplus; it could be halted at any time. When the money is exchanged for future investment such as a bigger boat, possibly even involving a loan, however, there comes a point of no return, where the only rational thing to do is to fish ever more to make the boat pay for itself, pay off the debt, accumulate more money to buy an even bigger boat for the capacity to sell yet more fish. This means that the more one gets involved in the logic of capital accumulation, investment and credit, the more difficult it gets to extricate oneself from the money economy's seductive pull. Moreover, once activated, the process seems to take on a life of its own, to run itself within the networked interconnectivity not of nature but of money.

On the basis of this simple story, we can also see the relation between surplus value and nature. Surplus value, argues Binswanger (1994:180–1), is first and foremost the outcome of the exploitation of nature – not as is still the favoured conception, a result of the exploitation of labour – increased efficiency or research and development. While those latter factors will have a role to play in the production of this surplus value, even they are rooted in the primary exploitation of nature. As a source of raw materials, nature is the ultimate root to material wealth. Production, according to Binswanger, is therefore the transformation of nature into money, into an abstract value that can be infinitely stored and accumulated outside nature's strictures of entropy and finitude. In *Beyond Beef*, Rifkin (1992) substantiates this analysis by establishing the link between the exploitation of nature through the production of cattle and the concept of capital.

> The word 'cattle' is derived from the words 'chattel' and 'capital'. Cattle are the oldest form of mobile wealth and have been used as a medium of exchange throughout much of Western culture. The devolution of cattle from the status of a divinity to the status of currency and commodities serves as a historical mirror to our own changing relationship to nature.
>
> (Rifkin, 1992:2–3)

> Even the Latin word for money, *pecunia*, comes from the word *pecus*,
> meaning cattle.
>
> (Rifkin, 1992:28)

Thus, he argues, the emergence of the Western cattle culture with its competition for and exploitation of the land is inseparably intertwined with the rise of capitalism.

This particular conception of nature = money, however, is built on the assumption that nature is free, that it is a resource without a price tag: as part of a limited earth environment, nature may be a scarce resource which, in turn, brings with it problems of allocation, but it is a resource that neo/classical economics considered external to the calculus of gains, losses and profit. This externalisation of nature brought with it the well-rehearsed problems of over-exploitation, and depletion of non-renewable resources. One economic response to environmental degradation arising from this economic tradition is therefore to pay for the natural resources by putting a price-tag on nature and calculating costs on the basis of cost-benefit analyses (CBAs). I discuss CBA and its implications for environmental praxis below. Here, I first want to consider a number of issues that bind the nature–money–time relation into a coherent whole. (On this subject see also Biervert and Held (eds) 1995.) To establish prices for all of nature's treasures means the superimposition of economics on nature. Similar to the application of Newtonian assumptions to life, however, this economic description of nature's goods and its prescription for environmental ills (caused in part by that economic approach in the first place) brings with it problems of its own. As I show below, economic assumptions, like their Newtonian counterparts, stand in a conflictual relation to the processes of nature. That is to say, the neglect of rhythmicity, seasonality, time-scale and intensity of change, the denial of the importance of timing and variability over the life-course, and the disregard for the centrality of reproduction and regeneration are just some of the pertinent time-based discrepancies arising from the application of abstract economics to the temporal realm of ecological processes. Moreover, the nature = money assumption and relation should not be seen in isolation from the time = money presupposition and approach to industrial activity. The two are subtly related and have thus to be understood with reference to each other.

## Time = money

Time = money is a presupposition that permeates industrial culture and most specifically the world of business: labour is paid by the hour, the week and the month. Companies calculate their labour costs in 'man-hours'. Surplus value and profit of a business cannot be established without reference to time. The life-span of a machine is reckoned in relation to the amount of work it produces within a specific period. Overtime, 'time out' through absenteeism, and strikes all form integral aspects of the calculation of a business' production costs, its efficiency and its performance in relation to competitors. The time = money equation and the associated use of time as a commercial medium for exchange are thus possible only on the basis of 'empty time', a time separated from content and context. Only as an abstract, standardised unit can time become a currency, a medium for exchange and a neutral value in the calculation of efficiency and profit. As such, the time = money equation has to be differentiated from the relationship to time as a resource.

To conceive of time as resource is to associate it with the inexorable passage of time and the inevitability of death. It is this boundedness of earthly existence, this birth–death framework, which turns attention away from the source(s) of life, away from *natura naturans* (the immanent creative and re/productive force of nature discussed in Chapter 1) and focuses instead on its limited supply. It means that awareness of earthly finitude and the irreversible directionality of life from birth to death makes time a resource that gets scarcer with advancing age. Similarly, the boundaries created by days and seasons shift attention towards the limited nature of this 'resource': only so much can be done in a day; equally, there are right and limited times for activities associated with the seasons, with planting, growing and harvesting, with celebrating Christmas and New Year. The birth–death parameter, the life stages in-between, seasonal and religious markers and the time bounded by night and day are not the primary consideration when time is associated with money. The resource rooted in boundaries is not what Benjamin Franklin had in mind when he wrote the by now famous lines, quoted in Max Weber's *The Protestant Ethic and the Spirit of Capitalism*.

> Remember, that *time* is money. He that can earn ten shillings a day by his labour, and goes abroad, or sits idle, one half of that day, though he spends but sixpence during his diversion or idleness, ought not to reckon *that* the only expense; he has really spent, or rather thrown away, five shillings besides.
> Remember, that *credit* is money. If a man lets his money lie in my hands after it is due, he gives me the interest, or so much as I can make of it during that time. This amounts to a considerable sum where a man has good and large credit, and makes good use of it.
> Remember, that money is of the prolific, generating nature. Money can beget money, and its offspring can beget more, and so on.
> (Franklin quoted in Weber, 1904–5/1989:48–9)

Conceived as money, time is no longer a resource associated with the transience and boundedness of life. The transformation of a lived resource that we can use, plan and allocate to one that is exchanged and controlled on the labour market has to be understood with respect to a very specific development: *the creation of a non-temporal time* and the orientation of social life to this very specific kind of time. The association with money, in other words, is only possible once time has become decontextualised and disembodied from events, once it has been established as an universally applicable, abstract, empty and neutral quantity that accords all hours the same value: where one hour and the twelve-hour day are the same regardless of whether the time in question refers to the morning or the night, in Newfoundland or South Africa, during summer or winter. This is a time where content is irrelevant. As money, time is irreducibly tied to work and economic exchange. Once it is quantified and used as an exchange value, time becomes an economic variable like labour, capital and machinery, a commodity that has to be handled economically: we can speak of a time economy.

This commodified time is to be distinguished from the lived resource by a number of key characteristics. Luhmann (1982b) identifies these as independence from context and content, mathematical expression and chronological dating, universal applicability and translatability. Crucial is also mental reversibility by which we are able to repeat what is fundamentally unidirectional and irreversible. Only in this decontextualised form, as Marx (1857/1973:140–43) has argued, can time become commodified, can it enable us to convert a variable quality into an invariable, abstract exchange value and mediate the value between different commodities. That is to say, only in this abstract form can time mediate the translation between fundamentally different qualities of the environment and cultural activity; only in this form can it be part of a cultural environment in which social relations have become context free and anonymous. Readers will recognise these features of commodified time as ones they have already encountered in Chapter 1 as key characteristics of Newtonian science. Through our relation with technology and commodified time the Newtonian conception of reality and its associated approaches to socio-natural life have seeped into the social fabric like a dye, subtly influencing every sphere of the industrial way of life. The contemporary link between time and money, therefore, has to be understood within a more general social context of increased distancing from the concrete and a shift from personal to abstract economic exchange. Not one aspect, moreover, but clusters of characteristics are implicated. At the very least, these involve the increasingly abstract nature of economic exchange based on de-materialisation, de-contextualisation and the empty, standardised time of calendars and clocks.

At the same time, however, we need to appreciate that time is not at all like money. There are significant differences that are worthy of our attention. Thus, for example, whilst time passes outside our control, money can be consumed at an intentional pace or it can be left to grow. Added days and years mean ageing, growing older and therefore closer to death and the end, while the accumulation of money means unrestricted growth of wealth, that is, growth that can continue indefinitely. As a constitutive dimension of our lives, time is lived, generated, and known. As a resource it is used and allocated. As a commodity it is exchanged for money on the labour market. Unlike money, however, time cannot be externally stored or accumulated. Whilst money stands in a direct quantitative relation to value – the more the better – this is clearly not the case with time, otherwise the time of prisoners would have to be accorded maximum value. Moreover, some time is more valuable than other time. Thus, the time of the professional far outweighs the money value of the unskilled labourer's time. Importantly, the money value of time is exclusively tied to paid work. Any time that falls outside this framework of evaluation is not exchangeable for money: the time of children and the elderly, the time of mothers and of those that care for spouses in the home, the time of prisoners and the unemployed – none of this time is money. Their time is 'valued' in the shadow of economic relations, filtered through that way of thinking about and evaluating the world. And since, furthermore, money relates to power, any time that is not readily translatable into money tends to be associated with a lack of power. This unsuitability to translation into money clearly includes the variable, contextual time of living processes and natural phenomena and this in turn gives us one indication about its low status. (See also Chapter 4. I have written extensively elsewhere on this important subject of the time–money–power relation and have gone there into far more depth than is possible in the context of this particular argument. For further detail please consult Adam, 1990: Chapter 5; 1993; 1994/5; 1995a: Chapter 4. Here, I merely want to indicate some of the critical features of this conjuncture of time and money for understanding economic approaches to time and their impact on the environment.)

Where time is equated with money, speed becomes an important economic value, since the faster a product can be produced, the less money-time is tied up in the process in the form of machinery, interest payments and labour costs. Speed increases profit and shows up positively in a country's Gross National Product (GNP). Here, the time = money intersects with the nature = money conception. In an economic context in which nature is external to the calculus of costs, gains and profits, the energy use associated with speed tends to be external to that calculation as well. The fact that speed is proportionally related to energy use is therefore kept out of the economic balancing act. When it is incorporated as an integral part of costing, however, a very different picture emerges: the faster something moves, the higher its use of energy, the higher the costs involved to both producer and environment. Instead of profit, the speed = energy equation features as a negative entry on the balance sheet.

The importance of speed in a context of time = money, however, relates not only to the process of production but also to speed of communication and the speed of innovation. Thus, with respect to the former, the British telephone company Mercury has used a big bill-board advertisement announcing that 'speed kills the competition'. The message is clear: you have to be faster than your rival to survive; even the gap (meaning disadvantage) of a few seconds can make a crucial difference to securing or losing a deal, be this in business or financial services. Nowhere is the importance of the speed of communication more evident than on the stock-market. Speed of communication, furthermore, is intimately tied to the changing fashions in organising the interface between products and consumers. Thus, for example, 'lean production' and 'just-in-time' manufacturing are means to reduce the time between manufacture of products and their delivery to the consumer, with the result that ever-growing numbers of half-empty vans and lorries criss-cross the country thus further increasing the environmental damage wrought by road transport and haulage. Equally consequential to the competitive climate of business is the speed of innovation: whoever brings a new product to the market first has a clear advantage over others who are in the process of developing goods and services along similar lines. With huge sums of money tied up in research and development, earliest possible production is crucial. I will return to this point in later chapters. With speed so tightly linked to efficiency and a competitive edge, however, we place a multiple interrelated burden on the environment: first with respect to a rise in energy use, second with the production of pollution, and third with reference to increased production of waste, all of which, in turn, add to an overall escalation of environmental degradation. A third pressure on the environment arises when, due to the economic valorisation of speed, corners are being cut in relation to safety and sustainability. Finally, an important indirect effect is created through the ever decreasing time for reflection, contemplation and deliberation on the processes and their effects.

Since the two assumptions, 'nature = money' and 'time = money', have such significant environmental impacts it is clearly important that we get to know them in order that we may recognise these naturalised assumptions as conventions, begin to build alternative conceptions and establish more appropriate practices. The difficulty of this endeavour, however, must not be underestimated. The roots to these assumptions and their associated approaches to socio-natural life run deep. Tied to the fear of impermanence, death and the unknown they connect with some of the deepest human desires.

### Encoded desires

Death is an inescapable condition of human life and as such it forms a central if implicate feature of our existence. Without knowing the time and place of death, our lives are lived with reference to it. Much of what we do as part of cultural life is geared towards coming to terms with finitude and/or trying to overcome it. All major religions engage with it. The creation of permanence in the context of ephemeral natural existence is one of the marks of culture: art and writing, temples and rituals, money and scientific laws are all ways of rising above the earthly condition of our physical being. Heidegger (1927/1980) was referring to this coming-to-terms with finitude when he proposed that transcendence is the essence of *Dasein* (contextualised Being/Existence). As concern, resolve, anticipation and projection towards our own possible future, Heidegger understood human existence as 'being ahead of itself'. To him, therefore, transcendence is the ultimate basis of all human knowing and acting. *Dasein*'s total penetration by birth and death is our ontological condition, our basis of existence. With it comes the encoded fear of non-existence, of nothingness after the end, and the desire for mastery and control over the inevitable. With the awareness of finitude and death, unmediated 'natural' living is no longer possible. It is in the relationship to finitude, to the birth–death parameter, therefore, that we have to find not just the human spirit but the characteristics of a culture and a specific era. Industrial culture is no exception: it is marked by a cluster of strategies for coming to terms with and overcoming the fear of nothingness and the unknown. With the creation of clock time, scientific laws, and a money economy it has de-temporalised the temporal, fixed the ephemeral, transformed transience into permanence, and thus created a simulacrum of immortality and omnipotence.

Intimately connected to the ontological condition of finitude is the inescapable temporality of being: the transience and ongoing rhythmicity of growth and decay, maturing and ageing; the continuous change at varying speeds and intensities; the eternal braid of beginnings and endings, pauses and rests. Like the spectre of death, temporality is a threat. It reminds us of the irreversibility of life and all its processes. It means loss of youth and beauty, the movement towards death, the past irrevocably gone and the future unknowable except for the certainty of the end – some time, some-where. Cultures abound in means to arrest those relentless processes, fix them for long enough to get to know and thus control them, repeat the unrepeat-able, construct permanence from transience. Technology, of course is central to this process of de-temporalisation with its associated control and so is the development of a money economy. It is worth our while, therefore, to look a bit closer at the desires and dreams encoded in the creation of both clock-time and money.

If the multiple time of our physical being is marked by finitude bounded by birth and death and by rhythmically organised transience then the creation of clock time has acted as a counter to those characteristics: it goes on indefinitely, day after day, year after year. It pulses to a metronomic beat, its motion precise and invariable. It is not subject to the natural processes of entropy, growth and decay. The time of the clock is quantified and standardised, unaffected by context and seasons. These features make clock time predictable, applicable in a uniform way irrespective of time and place. In its technological form, therefore, the ultimate threat to human control is tamed. A parallel metamorphosis has been achieved with the development of the money economy. When the produce of nature is exchanged against coins, that which is temporal and perishable is transformed into a permanent, atemporal value. In this form it can be stored and accumulated according to different principles. It has become subject to human control. The transformation of nature into money can thus be understood as the pursuit of alchemy. Embedded in that quest is the desire for control of the earthly conditions of existence, for unboundedness and permanence, for cheating entropy and death, for future security and certainty (Binswanger, 1991, 1994; Simmel, 1907/1978). Neo/classical economic relations strengthen not just the desire but the illusion.

## Economic habits of mind – the Newtonian heritage

'What', we may ask, 'is wrong with that pursuit if it underpins human culture, if it is a central part of that which makes us human?' The answer to this question is that the environmental degradation and the hazards we face today are linked to the scale and intensity of the impact of that pursuit as well as the extent of the time–space distantiation of its effects, its global dispersal and its reach to untold future generations. The achievements of the quest for transcendence and permanence are folding back upon themselves. Successes have turned into excesses and thus constitute an integral part of the contemporary environmental problems: they are devouring their bases of existence. Could it not be said that the greening of business and environmental economics are signs that these problems are being addressed? If concern for the environment and all things 'green' is becoming so widespread and if it is filtering through to the world of business and politics, is this not an indication that globally and locally we are getting to grips with environmental degradation and the manufacture of hazards, that we have changed course and are now moving in the right direction? Are we not beginning to put right what was the key problem in our relation to the environment by internalising nature both politically and economically, that is, by giving it a price tag and including it in the calculation of public costs and benefits, of capital outlay, expenditure, income and profit? When we bring time into the analysis, I am afraid, the picture becomes extremely complex and the response to these hopeful questions rather gloomy. The reasons for the negative assessment, however, differ significantly from those put forward by deep ecologists which so alienate those in politics and commerce who consider themselves to be dealing with 'the real world' and whose daily concern is money, 'the stuff that makes the world go round'.

There are two interrelated issues with respect to the habits of mind I am discussing in this and the first chapter: a) collectively, the linear perspective, Newtonian science, and neo/classical economics are a constituent part of and contributive factor to the creation of environmental hazards; and b) when applied as a solution they may offer temporary, short-term alleviation of the symptoms but, over the longer term, they pose the wrong questions, use the wrong conceptual tools, follow inappropriate paths. From a longer time perspective, therefore, they are likely to do more harm than good. Given their starting assumptions, this could not be otherwise. From a temporal perspective, the answer is not, however, to get rid of perspectives, science and economics; rather, it is to accept the need for re-tooling, to develop economic science assumptions and perspectives more appropriate to the tasks that arise from the serious long-term hazards and environmental degradations we are faced with today. A brief critical analysis of some assumptions associated with neo/classical environmental economics will illustrate my point. I discuss only a selection of such assumptions since neither this book nor this chapter is a tract in economics. The points I raise, therefore, are not intended to be exhaustive but merely illustrative of the wider analysis presented in this treatise. Thus, in the light of the overall argument about timescapes, I focus first on the prime economic practice of calculation and its associated dependence on quantification and measurement, on visible surface phenomena – the *Merkwelt* and *natura naturata* – and linear causality, on past-based knowledge and the pursuit of certainty. Since the vast range of environmental economic tools on offer today share the basic assumptions associated with quantification, the popular cost-benefit analysis (CBA) can serve as an exemplar for the wider range. Next I consider the troubling economic practice of discounting the future. Third, I attend to the issue of economic growth and distinguish its temporality from that of living processes and regeneration. Fourth, I discuss the potential of environmental economics for management, control and risk assessment. Finally, before I turn my attention to the invisible workings of market forces, I reflect on the image of spaceship earth and contemplate its un/suitability as an environmental metaphor.

### Calculation

At the simplest level, CBA is the comparison between costs and benefits of an action and/or its effect. With respect to the environment, CBA starts from the assumption that environmental action involves costs. The action might be pollution control or prevention, the development of cleaner technology or less harmful chemicals; it may be not felling a rain forest, re-routing a planned road or pipeline, re-siting an urban development. Costs are assumed to occur for both action and non-action, for development and lost opportunities since, as I argued above, the nature = money = time equation means we are dealing with resources whose non-exploitation automatically means a loss of money and opportunity costs. It is further presupposed that both costs and benefits can be established by asking people: ask the car manufacturer what it will cost to take out all the harmful substances from diesel fuel, ask asthma sufferers what it is worth for them to have the cleaner air. The comparison of costs with benefits, it is argued, allows us to make rational decisions about which actions make economic sense and which do not. The acronym BAT-NEEC which means 'best available technique not entailing excessive cost' arises from the CBA approach to environmental hazards and degradation and so does the practice of supplying pollution credits where heavy polluters can swap credits with 'underpolluters' so that the overall level set by government is not transgressed. (I come back to this issue in a later part of the chapter.) The benefits of CBA are considered self-evident: once we begin to start paying for the resource, i.e. nature, we not only treat it with more care and reverence but we also do our best (because it is in our self-interest) to avoid the pain of having to pay for the damage. Therefore, if this method is extensively applied, so the argument goes, environmental protection will be immeasurably enhanced. (For readers who want a more extensive introduction to environmental CBA, I recommend Jacobs, 1991: Chapters 6 and 16.) The extensive criticisms of this approach are well rehearsed and I will not repeat them at length. Since I am concerned here to critique some of the central assumptions that underpin economic measures like CBA rather than CBA as an environmental policy *per se*, a few examples of existing reproaches of this practice will suffice.

Critics have rightly pointed out that while it may be reasonably straightforward to establish the financial costs of pollution abatement to the polluter, it is far more tricky to calculate the benefits to the recipient. First there is the substantial difference between people being asked how much they are willing to pay for a clean 'resource' such as air or bathing water and being questioned about how much compensation they should be paid for the polluted air and water: no prize for guessing which of the two would better fit the BATNEEC principle, that is, the economically optimal path of degradation. Then there is the small issue that people's capacity to pay differs widely. Again the application of the BATNEEC principle results in the convenient conclusion that it is far more cost effective to pollute the environments of the poor. Next, there are a number of pertinent, if inconvenient, questions: 'What if the benefit is priceless?' 'What if the effects cannot be known due to the vast time-distantiation of the processes involved?' 'What if we cannot ask those affected because they live on the other side of the globe or are not even born yet?' And, finally, 'how do you establish equivalencies between death in degrees through radiation, asbestos or hormone-disrupting chemicals, the loss of ozone, your child crippled by asthma, beaches without faeces and oil globules, and the saving of a rare snail, a heathland, or an ancient peat-bog?' The credibility gap is pushed to its limits when we are asked to believe that the unitary framework of putting equivalent prices on all these costs and benefits may be difficult but not impossible, that it is merely a question of time and research funding before the resource, nature, is going to be financially mapped in its entirety. What optimism! From my discussions of the complex temporal features of environmental hazards and degradations in Chapter 1, it will be clear to the reader that I find this faith in economic prowess grossly misplaced. My lack of faith relates, on the one hand, to the inappropriate underlying assumptions which neo/classical economics shares with the linear perspective and Newtonian science and, on the other, to the fact that this type of economics knows no 'reality' outside the world of money. Not everything can be (nor should it be) encompassed in the economic and scientific fold. It is this otherworld, the shadow realm of economics, science and technology that I mean to celebrate and strengthen. First, however, let me use the focus on time to deconstruct the rationale of this neo/classical environmental economics.

If we take hormone-disrupting chemicals as our example, it will quickly become obvious that the dependence on calculation, quantification, and measurement and the practice of asking people to express their choices in monetary terms is not only impracticable but impossible. As soon as we are dealing with non-linear, non-proportional, time–space distantiated processes, neo/classical economics and other theoretical systems built on similar assumptions become meaningless. That is to say, with hazards such as hormone-disrupting chemicals, the economic system becomes inoperable long before we encounter the difficulty of asking embryos about the price they put on their future health and fertility; long before we could consider what price could possibly be put on the slow invisible destruction of the capacity to regenerate and reproduce life on earth; long before we could ask what price tag would be appropriate for the 'lost resource' of future generations of humans.

## Discounting the future

Not only does neo/classical economics have a problem with absent people who cannot express their choices in the market and with resources that are priceless, even more disconcerting is its relationship to the future. The future in this economic approach is discounted which means giving the future less value than the present. (For an excellent detailed economic treatise on this subject see Price, 1993; for a more general introduction see Jacobs, 1991.) This devaluation of the future makes perfect sense within a scheme that assumes that individuals act to maximise their self-interest. The rationale of discounting the future needs some further explanation. Given the choice, it is argued in favour of this practice that a person would prefer to have £1,000 in their hands today rather than in ten years' time. This money could be invested now and would therefore be worth much more in ten years than £1,000. Also, as incomes tend to rise over time, the person is likely to be earning more money by then, which makes the sum proportionally smaller. Finally, people do not know about their future need; they may even be dead ten years from now. The upshot of this is that economists discount the value of a future sum of money – cost, benefit, earnings – backwards towards the present. This means, by today's value and at a discount rate of 10 per cent per annum over a period of ten years, the future £1,000 is calculated to be worth a mere £386 today. The effect is that the future is devalued by a sleight of the economic hand. This approach to the future makes many an incomprehensible action rational – on neo/classical economic terms, that is. From the standpoint of the present, projected into the future and back again, the future is less important than the present and, given a long enough time span, it is in this scheme of things worthless. As the devaluation of the future increases with temporal distance – £1,000,000 of a hundred years hence is calculated to be worth a mere £75 today: a few more years and it is worth nothing. This logic is proving particularly troublesome when it is applied to the sort of long-term hazards I am writing about in this book. That is to say, from this economic perspective, the present of future generations who have to struggle with the legacy of hormone-disrupting chemicals, radiation, genetically modified organisms, and industrial farming practices is of zero value to us who get all the benefits from today's innovations and economic prosperity. Colin Price (1993) shows through careful analysis that discounting cannot be justified even on its own terms let alone with reference to future generations and the environment. He thus comes to the conclusion that the practice should be stopped.

> If each one of us who inclines to discount is obliged to explain to friends and family why the passing of time implies the decline of value, we may yet offer the future something better than our kindly concern: let us not scorn a cold and rational altruism, driven by a belief in the propriety of sharing with later times the things we have valued, in the time which has been given to us.
>
> (Price, 1993:347)

The practice of discounting of the future has to be placed alongside the assumption that as long as we keep investing some of the resource capital the monetary system will grow, and therefore future people ought to be better off and we need not worry about them as long as we have done our bit towards growth. The question is: 'what does this mean – better off on what terms?'

## Growth and regeneration

The growth of money tells us nothing about the state of the environment or future health. It cannot tell us anything about the conditions and bases of existence because it is an abstract system that operates without recourse and reference to the wellspring of its surplus value. The economy is traditionally conceived in terms of an endless circular flow of money: that is, from firm to workers, to goods and services back to firms, with accumulated surplus needing to be reinvested back into the system. This perpetual flow of money is the economic equivalent of Newtonian particles in motion: abstract, decontextualised, de-temporalised. In both perspectives, the physical

75

environment is external to, that is, excluded from the conception, and so is the irreversible, time–space unidirectionality of its processes.

While money may be endlessly circulating and growing, each transformation in the physical world involves a one-way direction of energy and information exchange where, once transformed, the resources can no longer be used for the same work in the next round of exchanges: the petrol used for driving a car is dissipated into the environment as pollution; after their use commodities end up as waste on the rubbish dump. While the economic exchange of money may be conceived as and may function in abstraction along Newtonian lines, the economy operates in socio-natural environments. It is always embedded and contextual, subject to entropy at the physical level as well as unequal demands, relations of power and econo-political (not just individual) interest at the social level. The difference between the circular flow of money and the spiral, directional change of socio-economic processes, as Michael Jacobs (1991) points out, is important since, with respect to the environment, it is not growth *per se* that is at issue but the growth of consumption and waste, pollution and hazards, together with the interconnections between these various kinds of growth. From a temporal perspective, furthermore, it means that the speed and intensification of these different kinds of growth as well as their time-scales and timings, the time–space distantiation of their impacts and the time lags between the causes of growth and their symptoms have effects. Each one of these temporal features matters. Each is implicated in the others.

A last crucial consideration with regard to growth is whether or not a system is given the chance to regenerate, not just itself but together with all its essential bases of existence, that is, whether it is sustainable. This, however, is seldom a question of money but of time and entails that we understand not just the entropic (thermodynamic) processes and dissipative structures of the physical world but also the creative temporalities of life. It means we need to encompass the regenerative growth processes that innovate, form new forms, and bring forth new species (on environmental reproduction see also Hofmeister, 1997). Growth not as the abstract flow of money but as a regenerative feature of life is inextricably tied to the basic material conditions of earthly existence, to sun, air, water and soil. Taking the sun as an example, we find that its radiant energy is a source of life and the relative movement of the earth–sun–moon constellation constitutive of nature's daily and seasonal rhythms which are deeply encoded in the very core of earth being: almost all life forms regulate their growth cycles of activity and rest to it. Our human physiology as well as our social processes are governed by it. Birds migrate and mate in relation to it and nearly all of what is understood as plant behaviour is linked to it. This temporality of life on earth is orchestrated into a symphony of rhythms of varying speeds, durations and intensities, of intricate timings and infinitely complex synchronisations, of reproductive cycles of birth and death, growth and decay. As such, the sun as source of life underpins everything else and is inextricably implicated in any growth and regeneration.

We tend to think of sun, air, water and soil as constant and continuously available 'resources', supplied and renewed through nature's own capacity to recycle and regenerate its own bases of existence – through, for example, living organisms, photosynthesis and processes of decay. With appropriate expansion of the time frame through which we understand these sources of life, however, not even the sun is a continuing resource: even this most stable and permanent element of our physical support system is itself subject to temporality. That is to say, whilst the sun is on one hand the source of earth time and its seasons, it is on the other hand itself historical, degradable, vulnerable to external cosmic impact, and not absolutely beyond influence from activities on earth. We need to appreciate, therefore, that the time frame of any understanding – the temporal perspective – is implicated in what we see: the temporalities of life – of change and rhythmicity, timing and tempo, speed and intensity, duration and succession – combine as a seamless whole with our temporal perspective. When industrial activity is beginning to threaten the earth's capacity for self-renewal, we need to think in terms of regeneration, not just of eco-systems but of the bases of existence. Currently, the industrial way of life has begun to affect the absorption and regenerative capacity of air, soil and water: the damage to the ozone layer, the loss of top soil, and the chemical pollution of the oceans and the ground water are pertinent cases in point. Thus, it is important that such a shift in understanding is accomplished and that it is accompanied by closer attention to how we conceive of environmental growth, regeneration and sustainability.

Now that economists are beginning to incorporate in their approaches notions of sustainability and re/production we need to be watchful that these concepts do not remain rooted in Newtonian science, that they go beyond notions of sustaining past and present conditions, beyond thinking of reproduction as repetition of the same. The re/generation of the bases of life and the life forms they support is never just a reproduction of what was before but entails fundamental renewal, genuine creativity. Each regeneration brings forth its own transcendence. With respect to sustainability, it is therefore not sufficient to think in terms of reproducing the past without simultaneously implicating in our concerns the innovative, open future. At the same time, however, we need to appreciate that, despite the creative openness of reproduction, it operates within limiting frameworks. Morphogenesis, the coming into being of form, for example, has shape, size and age encoded within a range, the final form unfolding and forming in interaction with the totality of its environment. Thus, an acorn will not grow to the height of 200 metres in the shape of a pine tree, just as a lamb does not grow into the shape and size of a camel. Even hurricanes and tornadoes form and develop within a flexible, yet bounded range of potential. (On the temporality of life see Adam, 1990, Chapter 3 and Luce, 1977; on morphogenesis see

Sheldrake, 1983.) This means, sustainable growth in and of eco-systems is reproductively creative and open with flexible limits to form, size and temporal expression. A second central feature is the fundamental interdependence and indivisibility associated with ecosystems and their bases of existence. Both the *natura naturans* of creative regeneration and its connectivity, however, are difficult to access from the materialist perspective of traditional economic science.

That is to say, such a perspective can only provide a surface vision of the phenomena in question. It objectifies and fixes what are largely immanent, interdependent, time–space distantiated, temporal, rhythmic process-systems. The boundaries of its reach are set by quantification: what cannot be quantified falls outside its jurisdiction, which means all things future, potential and contingent, invisible and immaterial. Quantification, moreover, depends on counting isolated bits: genes, trees, individual choices. Sun, air, water, soil, plants and animal species as the collective bases of existence, however, are indivisible. This indivisibility is multiple and finds expression in a number of ways: first, sun, air, water and soil are globally rather than nationally or locally constituted. Second, all of them, including plants and animals, are indivisible with regard to their interdependence. Third, with reference to their temporal extension, their long past of millions of years of transformations is inseparably enfolded in their current expressions just as their future potential is implicated in the here and now. This means that calculations such as those about the length of time it takes to grow ten centimetres of top soil, for example, can be a figure of speech only: there is no such thing as 'ten centimetres of topsoil' outside the connectivity and interdependence of the entire sun–air–water–soil–plant–animal eco-system. No matter how graphic, impressive and persuasive the abstraction, therefore, it ultimately misses the point. Equally, with all its past enfolded and its contextual relations with other systems, it is impossible for one tree to grow without the entirety of the eco-system being in place: the oxygen, sunlight, rain and seasons, the nutrients and micro-organisms in the soil, the insects and birds. The air carries volatile hormone-disrupting chemicals to the furthest reaches of our earth, and CFCs to the upper stratosphere; pollution enters the water cycle until it has penetrated to the deepest levels of the ground water: neither know national boundaries, neither can be quantified in a meaningful way.

And yet, of course, we have done just that. We have divided up and allocated the earth's air space and air waves, its rivers, coastal waters and oceans, its islands, land and mountain ranges, and thus turned collective bases of all-time existence into individualised commodities of and for the present. Treated as such they have been transformed into acculturated nature to a point where it is meaningless, as I argued in Chapter 1, to persist with the nature–culture dualism and, by the same token, pointless to persist with the environment–economy antinomy. Instead, as I argued previously, it is important that we understand differences and distinctions while simultaneously recognising the mutual interpenetration and reformation of socio-natural, glocal (global/local), economental (economic/environmental) processes and phenomena. After some 200+ years of thinking in terms of dualisms and isolated parts in motion, however, the habit is well engrained and difficult to transcend – but useless nevertheless for understanding ourselves in and of living nature. We have to transcend the habit therefore and appreciate to the depth of our being that the damage is not to one tree, not to 2000 birds caught up in an oil slick, not even to an entire species of whales. Rather, the health of and damage to one being or species affects the whole, and the entirety is implicated in and central to the well-being of every other member being.

Just as the conceptual habits of abstraction, fragmentation and thinking in opposites create false impressions, so the custom of thinking of time in the singular as a measure of physical and socio-economic processes misses the point completely: the hologram, as I argued in Chapter 1, is therefore a more appropriate metaphor for environmental engagement than a machine that can be dis/assembled from and into component parts. And with respect to the temporality of life it is the indivisible rhythmicity of improvised jazz rather than the infinitely divisible artefactual time of the clock that provides the more resonant imagery. Since imagery has such a powerful influence on perception we need to look briefly at one of the most pernicious metaphors of environmental economics – the spaceship earth – without which the economic perspective on the environment would be incomplete.

### Spaceship earth

Spaceship earth is to me the ultimate metaphor for Newtonian environmental economic science. It entails an image and assumption of the earth as a *man*-made machine: technical therefore knowable, subject to techno-scientific control, physically and spatially bounded, constructed from separate, interchangeable parts – if one part breaks it can be taken out and substituted with a new part – and thus subject to technical and technological fixes. Spaceship earth may be immensely complex but in the form of an artefact it becomes intimately and principally knowable; its actions and reactions become ultimately predictable. Moreover, when things go wrong it is possible for scientists and engineers to work out why this might be so: in human creations the cause-and-effect links are traceable through the system in both directions, that is, forwards and backwards in time. Finally, spaceship earth is a bounded system, encased (in metal) and hermetically sealed. This means, on the one hand, that it is safeguarded from outside atmospheric incursions and, on the other, that its internal resources are limited and finite, clearly definable, quantifiable and measurable. Given the issues I raise in this book, I cannot think of a worse and less suitable metaphor for the earth environment.

Everything about it is inappropriate. Every tiniest aspect of the image carries the wrong message.

The notion of earth as a bounded closed system of finite internal resources – a notion shared by Newtonian science (with the exception of the extremes of the time scale of existence, i.e. the beginning and the end), Darwinian biology and by classical geography and geology, and, of course, by environmental economics – needs some further attention. Boundedness, stability and internal linear, progressive change tend to go together and these knowledge principles are inclined to be repeated as we move through the levels of being from earth to cell system – everywhere images and assumptions of neatly functioning machines. In all these expressions temporal time is excluded since closed systems negate the immanent time of change and transience, of contingency and potential, of entropy and creative regeneration. The source of temporal time, we need to appreciate, is open-ended interaction and asymmetry, the spiralling cycle rather than the circle, the exchange and transaction, the give and take that constitutes life. Interaction, moreover, does not cease with the earth's atmospheric outer layer. There is no closure since earth is inseparably dependent on the sun's energy and the sun–earth–moon system is in turn tied to the solar system and from there indefinitely to the furthest reaches of the universe. The boundaries, in other words, are arbitrary and are relative to the applied theory and perspective. Contemporary earth science, for example, has moved from a closed- to an open-system understanding where the earth's non-linear history is theorised with reference to celestial influence and the impact of comets as a major creative/destructive force (see Davies, 1996). An important point to appreciate here is that from a temporal perspective closed systems are dead systems. For there to be life and change, systems have to interact, transact and exchange as integral aspects of wider wholes: they have to be open and their processes contextual and contingent. This, of course, is a pre-condition for life not only in the physical world but in human society in general and economic activity in particular. Not surprisingly, with the traditional conceptual tools discussed in this and the last chapter, it is difficult to engage in a meaningful and appropriate way with the central features of the re/generative force of contemporary existence and its hazards. Worse still, the tools facilitate the illusion of control and the belief that human creations ought to be amenable to management, be this of an economic, a political, or a scientific kind.

### Management, control and risk assessment

Integral to the neo/classical environmental economy of quantification, commodification, externalisation, and de-temporalisation is the assumption that control is possible, that sound economic science facilitates sound management. With the use of appropriate economic instruments, it is assumed, risks can be assessed and environmental problems managed – at the local level by business, science and local authorities, at the global level by science and institutions such as the world bank. These pre/suppositions are worthy of some attention. Let us begin by looking at the notion of environmental management and consider some of the parameters for its successful achievement.

The capacity to manage is dependent upon a number of essential preconditions. Central among these is the boundedness of that which is to be managed, that phenomena and their effects are delimited not just in space but also in time, that they occur in a known place and have a discernible beginning and end — note the link here to the spaceship earth metaphor. Equally fundamental is a reliance on the ability to establish causal connections and identify causal chains of events, which means, unambiguous relationships across time and space. Third, and closely allied, is the expectancy that cause and effect are proportional, that small causes have small effects and big events have proportionally large impacts. A fourth crucial prerequisite to management is the accessibility to measurement, quantification and control. Finally, 'solutions' are constructed on the basis of a known past projected into the future. As the chapters in this book amply demonstrate, environmental hazards put these presuppositions into question. Whether we are encountering the impact of synthetic chemicals, ozone depletion, air and water pollution, radiation, or a new disease such as BSE, the defining features seem to be spatio-temporal unboundedness, non-proportionality, time–space distantiation, contingency, and a high level of indeterminacy. Industrially produced and induced environmental hazards and degradations tend to be characterised by invisibility and periods of latency after which outcomes are no longer traceable with certainty to original sources. Often problems are only recognisable as such after they have been identified through the mediating loop of science and once they have been brought to public attention through media representation.

The socio-environmental conditions are at odds, therefore, with the assumptions and practices associated with the economic management of the environment: first, ameliorative action tends to be focused on visible symptoms only, and second, clock-time and linearity tend to exhaust the range of temporal facets. Causes, however, cannot be established on the basis of traditional, materialist scientific reasoning and evidence. That is to say, time-lags, latency periods and broken event chains as well as the gap between perception and impact, between *Merk-* and *Wirkwelt*, transform the pursuit of proof based on backward causation into an impossible dream. Since the interaction of rational technological systems with open, generative ecological processes creates inescapable indeterminacy, the future cannot be managed on the bases of past experiences. At the same time, however, environmental hazards are always symptoms of past actions which require responses.

With the broken causal link to the past and the indeterminacy of possible effects in the future, the potential for action in the present seems not to be captured by what is conventionally understood as management. In fact, it makes the idea of environmental management a contradiction in terms. Moreover, since effects cannot be delimited, that is, since they extend potentially globally into an open-ended future, it is difficult if not impossible to establish meaningful objectives and targets. And finally, since effects are not proportional to their causes and are not distributed in a uniform way, they need to be tackled in an eco-holistic way, which means with reference to both interactive connectivity and specific contexts. Thus, the conjuncture of contingency, contextuality, indeterminacy, latency, openness towards the future and non-proportionality renders the concept of environmental management highly problematic. We tend to be able to ignore the problem as long as we conceive the environment in spatial terms. That is to say, the environment as physical space, as a global realm of dwelling, gives the illusion of conventional manageability and control. As soon as we re-focus and re-conceptualise the issues in temporal terms as a *Wirkwelt*, however, what had been previously unnoticed becomes obvious.

A similar situation arises with reference to the calculation of risk. Like management and control, risk assessment is central to both economic and scientific practices. When the BSE crisis broke in Britain, journalists on behalf of the 'public' insisted on a quantification of the risks attached to eating beef. From an economic perspective it is not only rational but essential that risks are calculated against costs. This, as I show in Chapters 4 to 7, proves impossible in situations where causes and symptoms are separated by an open, unspecifiable time gap of invisible impact, a gap between the perception of symptoms (the *Merkwelt*) and the in/direct, non/linear, non/proportional, invisible production of the time–space distantiated impacts (the *Wirkwelt*). The language of risk assessment and calculation becomes inappropriate in process-phenomena that are time–space distantiated, contingent, interconnected. Environmental degradation and hazards pose threats that differ from the calculable risks associated with car accidents, thefts and house fires, in that they are neither just unintended consequences of rational actions, nor mere side effects, but *endemic* to the scientific innovations and economic practices characteristic of the contemporary industrial way of life: for hazards arising from the industrial way of life, the past gives no guidance for the future, provides no basis upon which to calculate and quantify risk. The lack of knowledge and certainty about the hazards, therefore, should not be considered as temporary uncertainty but as fundamental and inescapable indeterminacy (see also Beck, 1992a, 1996; Giddens, 1994a; Wynne, 1992, 1996). Indeed, insofar as risk implies calculability and/or individual as well as socio-economic decisions based on choice, the language of risk is inappropriate for time–space distantiated environmental hazards. The way the two terms are used in everyday language gives an insight into the difference: one takes a risk which implies calculation and choice but one does not take a hazard; instead, one faces hazards or is threatened by them which implies lack of control and an absence of choice. Business takes risks which turn to no-choice hazards for the wider public.

If the age of Newtonian science and its cognate neo/classical economics was marked by optimism and a trust in the human powers of calculation, prediction and control, knowledge associated with the industrial way of life of the late twentieth century is no longer characterised by such confidence. The spectre of indeterminacy, the unsettling of much that had been taken for granted, and the sense that the products of the industrial way of life are out of control have tempered the faith in those traditions and confront us with problems that need tackling from outside the classical framework of thought. Ulrich Beck designates societies at the end of the twentieth century 'risk societies'. What he encompasses within that term, however, is the condition for a 'hazard society', a collective future that falls outside the traditional economic way of dealing with the unknown, outside the capacity to calculate risks and insure against them. In other words, the globalised effects of radiation, genetically modified organisms, and hormone-disrupting chemicals, for example, are not insurable and neither are the consequences of global warming, ozone depletion or acid rain. These environmental threats are uninsurable, non-calculable and exclude the element of choice. With the emergence of these hazards, therefore, we have left the relative safety of the world of risks and entered the realm of hazards. To persist with the language of risk for analyses of such environmental hazards is thus inadvertently to contribute to the illusion of calculability and control, to nurture the hope that the knowledge traditions of neo/classical economics and Newtonian science will come up with tried and trusted solutions. That is to say, it encourages 'business as usual'. It is therefore a serious hindrance to the urgent task of thinking about these new problems in new terms.

## Market forces and regulations: encoded realities, carrots and sticks

Talk of fields and forces invariably refers to an invisible, encoded reality. The economic sphere is no exception. In market forces, the economy has its own version of a hidden reality. Not a thing, not a person and not an institution, market forces act outside the control of people and economic management yet are simultaneously involved in producers' decisions through pressures arising from its symptoms. That is to say, something other than individual economic choices facilitates the development of hormone-disrupting chemicals, hazardous technologies, and the production of increasingly toxic waste and pollution. Concern with these invisible market forces is important because it allows us to see how environmental degradation and the manufacture of hazards may be rooted simultaneously in the unintended consequences of (desired) individual actions and in the economic system as a whole, where it is an integral part of the invisible working of the market, the

*Wirkwelt* of market forces. Time is once more crucially implicated in the process and its symptoms. To engage seriously with the problems, therefore, means to focus on that which is normally left unattended, which means transcending conventional conceptual tools and habituated vision.

In order to see how this particular encoded reality is implicated in the production of hazards, we first need to consider the nature and workings of market forces. Market forces, Michael Jacobs (1991:23) insists, need to be strictly defined and differentiated from the coverall concept of 'the market'.

> Market forces come into being, at the level of the whole economy, when the majority of decisions made at the level of individual firms and consumers take place in markets, and there is no one determining the collective consequences. Market forces are thus the overall sum of many millions of separate individual parts. . . .
>
> Market forces are a macroeconomic phenomenon, operating at the level of the economy as a whole. Markets by contrast are a microeconomic mechanism, operating at the level of individual products, businesses and households. They are defined by the existence of a number of suppliers and purchasers free to choose with whom they trade, where changes in price mediate between supply and demand.
> (Jacobs, 1991:24)

It is on this basis that markets can be regulated and market forces, which are both supra-national and time–space extended, cannot. While the market can, to some extent at least, be understood through the Newtonian perspective of neo/classical economic science, market forces fall outside this vision, elude its grasp.

The key to both the indeterminable nature of market forces and their negative influence on the environment seems to be tied to two interrelated facts: first, that individual decisions and actions are not made with reference to each other, and second, that actions and their effects are time–space distantiated and thus passed on to third parties. With respect to the first point this means that private practices are not explicitly networked in a wider frame of common interest: no glocal society of earth-air breathers, no close-knit community of soil guardians, no association for the purity of the earth's waters, no glocal lobby against genetic engineering, no pact amongst the world's producers of hormone-disrupting chemicals to cease the production of these and other persistent synthesised chemicals, no resolution by the mothers of this world to eradicate plastics from their lives. Garrett Harding (1968/1992) coined the phrase 'the tragedy of the commons' for this phenomenon. He suggests that freedom in a commons brings ruin to all since, as independent, rational, free enterprise agents, we end up fouling our own nest, consuming and polluting our way to extinction. Harding cites as his example a piece of communal grazing land. This common asset can support only a strictly limited number of animals before the 'resource' is overgrazed and thus deteriorates to a point where everyone loses out. From the collective's perspective, therefore, it makes sense to keep to the optimal number of grazing animals that maintain the health and reproductive capacity of the land. From the perspective of the individuals, however, it is 'rational' (in the short term, it has to be said) to maximise their numbers of animals feeding on that free resource since this brings the highest profit to them now – in the present and the immediate/short-term future – at minimum expense, since the costs of this 'selfish' action are going to be carried equally by all members of the collectivity and are not to be paid for until some time in the future. The individual's profit, moreover, can be invested and thus increase personal wealth. Combined with the practice of discounting the future, there is no question what constitutes 'rational economic action' for the individual: concern for the collective good makes no economic sense; it is simply irrational. The logic also works the other way around: I might as well take the car for this journey since my taking the bus makes no difference as long as everyone else is still enjoying the convenience of this form of transport. Many environmentally damaging actions, when examined on these terms, turn out to be rooted in the operation of this economic principle and the *Wirkwelt* of market forces. 'The tragedy of the commons' is therefore one economic explanation for the rationality of polluting and environmentally destructive activities in a context where everyone knows that it is harmful and potentially even devastating to all of us in the long term.

The second point goes beyond the difference between the short- and long-term gains to both the time–space distantiation and the time lag between action and symptom. Thus, a key characteristic of the working of market forces is that actors rarely face the consequences of their actions. That is to say, perpetrators of an action or a practice tend not to be confronted with the effects because the symptoms arising from their activities surface at some other time in another place – downstream or to the West of the hub of polluting activity, in the depth of the oceans, in the coastal regions of a country on the 'other side' of the globe, in future adults of all mammal species. In economic terminology, the problem is externalised. We have, of course, encountered externalisation before at both the technological and conceptual level. We have come across it with the development of the clock, the linear perspective, Newtonian physics: all powerful and effective externalisers of time, reality and nature; all successful at distancing the cultural self from the contextual physicality of its being/becoming; all efficient objectifiers and with it negators of responsibility. *Hazards thus externalised become detached, free-floating, independent agents belonging to everyone and no one in particular.* For such externalised hazards, therefore, it is exceedingly difficult to retain a sense of ownership and responsibility and it is certainly impossible to apply the Polluter Pays Principle. One economic resolution to the collective damage wrought by the working of market forces through externalisation of effects and the economic rationality of individual action is to impose collective curbs on the rationale of personal/company gain. Recognition for the need to impose such checks is tied to the appreciation that change has to be achieved not only at the individual but also the collective overall level of environmental damage.

The instruments available to impose such curbs are many and diverse. They can be of an economic, a legal and a political nature; most measures, however, are based on a combination of these. Not their detailed description and assessment but their underlying assumptions and their relation to time(s) is going to be my focus of attention. When it became clear that the Polluter Pays Principle is inoperable with respect not only to establishing a link between polluter and damage but also to quantifying the costs, the pragmatic solution was to shift emphasis from damage cost to avoidance cost. This rescued both the principle and the economic reasoning that underpinned it. Where even this is considered unreasonable for the individuals and firms concerned, the Community Pays Principle is activated and the public called upon to bear the cost (Weizsäcker, 1994 Chapter 10).

For analysts with a strong commitment to economic principles and ideals, it is important that the collective measures are designed in such a way that they do not interfere at the micro level in the successful operation of the market, yet effectively curb the excesses of the market forces. For Michael Jacobs (1991:125) this means, 'choosing the macroeconomic *outcomes* of economic activity, not the laying down of its microeconomic *methods*'. He wants this approach differentiated from the traditional one which took a far more individualistic stance where specific kinds of pollution such as car exhaust emissions were targeted at source rather than with reference to the overall level of pollution, degradation or hazard. The difference, he argues, is crucial since fitting of catalytic converters to cars, for example, will not reduce the overall level of pollution as long as the numbers of cars and their use keep rising. When the focus of attention is on the overall problem, the understanding of the causes will change and the proposed solutions will consequently differ substantially from those that seek to tackle the problem at source. On the face of it, this approach seems to be the only meaningful way to reduce overall pollution, environmental degradation and production of hazards.

Whilst emphasis on macroeconomic outcomes makes eminent sense at one level, when combined with economic reasoning it brings forth such bizarre regulations as pollution credits which are attached to the US Clean Air Act of 1990 and more recent legislation on water purity. Like milk quota, these credits can be bought and sold on the market. There is, however, a difference: pollution is not like milk. While the principle makes sense at a quantitative level it fails miserably when we acknowledge the qualitative and temporal dissimilarity between the two 'products'. Under such a scheme, everyone is given an upper ceiling for pollution which means companies that are heavy polluters have to buy extra credits, whilst low polluters can sell theirs and thus make money on the basis of their 'virtue'. The argument is that the rational economic response is to pollute less and less because it costs money. So far so good. But what are we to make of the notion of *underpollution*, and where is the inequality in wealth between nations to come into this? Economists argue that this measure works in the favour of Third World countries: they can swap their pollution credits for hard currency! They desperately need money to pay off their debts to the First World so, the argument goes, it is extremely fortunate for them that they as underpollutors can now exchange their debts for pollution. What they get along with the cash, as integral part of the package, however, is not 'goods' but 'bads' to use Ulrich Beck's terminology. From an economic concern with overall pollution, it can become perfectly 'rational' to consider some parts of the world underpolluted and consequently argue for the prudence of spreading pollution more evenly across the globe. And this is not the only problem I have with pollution credits. What if the implicit message carried by the principle of pollution control at the level of macroeconomic outcomes is *not* to be frugal with pollution but rather that pollution has become a *right*, that is, everyone's basic right to pollute, a right upon which money, even fortunes, can be made, a right upon which political leaders can trade their citizens' physical well-being, and the basis of existence of future generations of humans and other species?

Since the mid-1980s member countries of the Organisation for Economic Cooperation and Development (OECD) and the European Union (EU) have made a strong commitment to the application of economic and regulatory instruments for environmental protection. Weizsäcker (1994:124) lists the following as principle measures with specific examples of the way they are applied in OECD and EU countries: emission charges for pollution of air, water and soil; user charges for municipal waste disposal; product charges for such things as non-renewable bottles, batteries and lubricants; administrative charges for such processes as registering genetically modified organisms or new chemicals; tradable permits for emissions (discussed above); extended environmental liability which means the duty to have adequate liability insurance; and deposit schemes such as those already operating for newspapers, bottles, plastic and cans. Most of these measures, of course, are still of the more traditional kind that considers that having price tags for individual environmental goods, services and damages is the appropriate solution to environmental problems. How such measures are applied, in turn, differs between countries and their interpretation of the task; in other words, whether they perceive environmental control as regulation of the market or market forces, and whether it is to be achieved by financial and/or political–legal means. Globally, the overall effect of the measures will be correspondingly uneven.

Whatever the method chosen, however, step one would always be to establish environmental indicators of current levels of pollution, degradation and hazards and on the basis of these to define targets and time scales within which they are to be met. The next step is to design economic, political and legal instruments by which these targets are to be accomplished. This is an excellent strategy for processes that are fully known in their development, are time–space coupled, linear in their causal progression, and proportional. Where this cluster of pre-conditions does not apply, however, these instruments and measures become less convincing: indicators have to be based on current measurements, but time-lags and indeterminacy mean that the measured levels of current degradation, pollution and hazards may not reach the height of their effects until ten, fifty, one hundred, even one thousand years hence. Trans-generational, invisible, additive and cumulative effects are rarely studied and tested for in their full temporal extension – how could they be? In such circumstances, where and when are 'acceptable levels' to be pegged? Where and when are definitions of safety to be defined? Where, when, over what time scale and at what speed are targets and their achievements to be set? As Colborn et al. (1996) argue with reference to hormone-disrupting chemicals, the past and the present do not tell us about the future-extended current condition of potential harm.

> Like generals, pesticide regulators are always and perhaps inevitably fighting the last war. Again and again, they have vetted chemicals for the most recently recognized hazard only to be blindsided by dangers they never thought to anticipate.
>
> (Colborn et al., 1996:241–2)

> The dilemma is simply stated: the Earth did not come with a blueprint or an instruction book. When we conduct experiments on a global scale by releasing billions of pounds of synthetic chemicals, we are tinkering with immensely complex systems that we will never fully comprehend. If there is a lesson in the ozone hole and our experience with hormone-disrupting chemicals, it is this: as we speed toward the future, we are flying blind.
>
> (Colborn et al., 1996:243)

Over the last fifty years, hormone-disrupting chemicals, for example, have become so pervasive in the environment and our bodies that there is no longer a norm against which danger and damage could be measured: from Inuits living a traditional life in the Arctic to beggars on the streets of Delhi, no one has escaped this particular contamination. With 1000 new chemicals entering the market every year (Colborn et al., 1996:137), what are the chances of achieving meaningful measurements and indicators? Surely, in such circumstances sensible targets and regulations have to be developed outside the tradition of measurement and calculation, just as discussions have to shift from the search for substitutes to questioning about the principles upon which both product and process are pursued. This does entail the need for an overview of a particular problem but it does not mean that the tools of economics, science and technology can provide the answers; rather, it is a question of values and, dare I say it, it is *wisdom* that has to set the goals. Only as a secondary step is there a need to worry about the economic, scientific and technological means to get there. Thus, if the vast numbers of chemicals pose a problem to monitoring and therefore to safety, then one target should be to reduce their number to a level that can be properly monitored. If, as is now widely accepted, the problem is not just toxicity but persistence and the fact that many of these chemicals do not break down into components that are easily absorbed by nature, then a second aim should be that this small number should be made up of chemicals that are non-toxic and break down quickly in a way that is compatible with nature's own chemical processes. Colborn *et al.* (1996:226–7) report on the efforts of Michael Braungart and William McDonagh who have developed criteria and guidelines that show the way to a reduction in both hazard and waste and thus provide a means for the better tracking and recycling of chemicals:

- Greatly reduce the number of chemicals on the market;
- Reduce the number of chemicals in any given product;
- Make and market only chemicals that can be readily detected;
- Restrict production only to products that have a completely defined chemical make-up and stop production of products containing unpredictable mixtures of chemicals;
- Do not use a chemical unless its degradation in the environment is well understood. (Reported in Colborn *et al.*, 1996:227)

There are already encouraging developments from business, as I indicate below. They are few, but with simultaneous encouragement from appropriate regulations their lead could show the way for others. Regulations cannot and do not happen in a vacuum. For them to work they have to be developed in a context of openness to the principles they seek to promote. In most cases, this is an interactive process with some businesses and some scientists leading the way, against the tide, already indicating the necessary change of direction.

For economists who subscribe to a Hobbesian view of human nature – that people are selfish and rationally pursue their own best interest unless curbed by socio-political regulations – there is an inescapable need for economic, political and legal instruments to counter human nature and market forces. This view, however, leaves unexplained a vast number of developments from within and outside the time-economy of environmental relations. It is these I want to turn to next, starting with a brief look at a variety of responses to the environmental challenge from business.

## Sunrise business – moonlighting for the environment

Above I have indicated the problematic relationship between the Newtonian assumptions of traditional economic science and the socio-technological context within which they are applied. I have argued that the assumptions, the conceptual tools and the instruments of neo/classical economics are unsuitable to engage in a meaningful way with the contemporary industrial way of life and the hazards that arise from it. Applied to environmental economics, this means that as long as its proponents adhere to those basic assumptions they are fighting a previous war or, to use a different metaphor, waiting for a train at a station that has been closed down years ago. I have been working for over a decade to expose the pervasiveness of the Newtonian world view and its counter-productive effect on almost every sphere of our lives here and now as well as globally and for the future. I am not alone in this endeavour. In each subject field there are colleagues who have come to corresponding conclusions and are similarly engaged in destabilising the certainties of their disciplines. Christine Busch-Lüty, a German ecological economist, is one of these persons. Over the past ten years, she has been pursuing arguments that overlap with the ones I have presented here and in previous work, suggesting that the single perspective, atemporal, decontextualised and absolutist approach of Newtonian economics is incapable of embracing the principles of sustainability and the uncertainty inherent in the contemporary environmental crisis. She insists that the traditional approach needs to be opened out to encompass not just the money economy but the physical economy as *oikos* (the whole house/hold), not just a market but a self-sufficiency based economy, not just production but reproduction, not just paid employment but non-paid and self-governed work, not just utility and selfishness but altruism, not just competition but cooperation, not just individualist but collaborative and collective perspectives, not just questions of money but value, not just material but immaterial, spiritual, ethical and emotional concerns (Busch-Lüty, 1994; Busch-Lüty and Dürr, 1993; also Biervert and Held, 1994, 1996; Biesecker, 1994).

In this part of the chapter I want to show where and how such approaches are practised or at least developing within and outside the traditional economy, thus providing, in parts, evidence that counters the economic view of human nature and social organisation. I was interested to note that companies who have taken on board the environmental message and consequently pursue sustainable business goals conceive of their enterprises as *sunrise businesses*: on the up, filled with the energy of the sun, bright and leading lights, bringing life and enlightenment to the darkness of a doomed way of doing business. The imagery is fascinating particularly since Carl Gustav Jung has worked with the sun and moon respectively as metaphors for the male and female principles. The language he attaches to the male sun is very much that of the sunrise businesses while reflection, soft light, darkness and invisibility and, of course, dependence on the primary force are central to the female moon. While it is my intention neither to get involved in a critique of Jung's work on this topic, nor to distinguish perspectives and assumptions along strict gender lines, I do want to make use of those metaphors, playing on their distinctions, differences and interdependencies as an illustration of the issues I want to raise here.

## How green is my business?

The world of business, and particularly large companies, has moved well beyond what environmental economists and politicians give them credit for: green is the 'in' colour, 'biological' a favourite term, the environment a central feature of advertising. 'Green', 'sunrise' and 'the future' have been conjoined to signify the direction for the forward-looking, politically correct business. After the preparatory conference for Rio in May 1990 in Bergen, the conference members asked Stephan Schmidheiny to prepare industry's input to the 1992 United Nations Conference on Environment and Development (UNCED). He brought together leaders from over fifty top businesses from across the world and formed the Business Council for Sustainable Development (BCSD). The report (Schmidheiny *et al.*, 1992), which was prepared for the Rio Summit, contained not only encouraging examples from across the world but also gave guidance on how to change practices in order that sustainable approaches to business could become commercially viable (see also Weizsäcker, 1994:146–9). A widely cited example of the commercial success of sustainable business is the 3M company of Minneapolis, a major US office machinery company. They changed the three PPP's of the Polluter Pays Principle to Pollution Prevention Pays and demonstrated that a rigorous re-use of chemicals can save a company money – in their case, $300 million over a period of ten years. In the meantime, not only pamphlets but books on the subject proliferate, giving every kind of advice from how to do an environmental audit in your company to checklists for your vehicle and transport policy. Moreover, it seems that most of the larger companies, including such well known ones as Shell, Tarmac, and Volkswagen have published statements on their environmental policy. Today, green is fashionable to a point where 'we are all green now' and where green can mean almost anything. 'The recent development of corporate environmentalism is proof, if any were needed', argues Ross (1994:4), 'that nature is the ultimate people-pleaser, whose name has been lent to and honoured by causes associated with its destruction'. It has become pertinent, therefore, to begin to make distinctions, to differentiate between shades of green: between greenness in business practice, product and process, perspective and approach; between the greenness of the technology and the chemistry involved. Again, I shall neither survey the entire field nor offer a critique of the multitude of practices. Instead, I focus on underlying assumptions

91

and on whether and what kinds of time are incorporated in particular approaches.

There are significant differences in the shades of green when it comes to choosing between technologies. Historically, technology has undergone radical changes in public perception. From being endowed with almost magical powers and the capacity to create cornucopian futures during the first part of this century it turned into environmental enemy number one during the 1970s when nuclear, chemical, and oil-based technologies began to get 'bad press'. During the 1980s, finally, we saw the rise of 'green technology' – the use of technology for the alleviation of problems rooted in the industrial, technological way of life. In the context of what I have said in the first and this chapter, what could the concept of 'green technology' possibly mean? (For an extended discussion on this topic, see Adam and Kütting, 1995.)

The proponents of 'green', 'clean' or 'environmental' technology (these terms seem to be used interchangeably) differentiate between reactive and proactive kinds. Reactive environmental technology, as the name suggests, is developed in response to a problem: it is added on, fitted to existing installations. In the case of pollution, therefore, it would lessen emission output through such add-on technologies as gas desulphurisation plants, catalytic converters, and air filters. What is less often focused on, however, is the fact that the pollutants are only collected, which means they still need to be disposed of and thus still end up as a burden to the environment: the problem is simply shifted from one medium to the other, from the air to the soil, for example. Moreover, as Sonja Böhmer-Christiansen and Jim Skea (1991:45) point out, 'for every 100 tonnes of coal burned in a power station fitted with FGD [flue gas desulphurisation], approximately 6 tonnes of limestone are required and 10 tonnes of gypsum are produced' which, in turn, leads to further environmental problems (see also Gourlay, 1992:178–9). The second category of environmental technology encompasses technology that works on a preventative basis as, for example, in the production of lower levels of emissions. As such, it generates less pollution than more conventional technology. 'Clean technology' not only saves on disposal costs for pollutants but it also tends to be more energy efficient and thus have lower running costs. Examples of the proactive type of 'green technology' are new types of power stations and light bulbs, windmills and solar panels.

Even before we can re-enter the debate about the technology–eco-system difference, however, we briefly need to consider the varied ways we can conceive of pollution, since this difference too has a significant effect on what one might understand under the terms 'green technology' and 'green business'. If pollution is defined as *effect* – the definition favoured by the UK – then the establishment of pollution is dependent upon scientific evidence and proof of causal connections. Here technology is needed to measure and establish harm. Yet, as Böhmer-Christiansen and Skea (1991:50) rightly point out with reference to their study on acid rain policies, 'establishing causal links between emission and environmental damage is not a matter of unambiguous deduction from known facts, but one of inference, a more elusive and subjective process. 'Direct evidence' for acid rain damage is, therefore, essentially impossible to obtain'. Time–space distantiation, time lags between cause and symptoms, and fundamental indeterminacy, as I have argued above, will ensure that 'proof' and unambiguous 'facts' are unobtainable. If, in contrast, pollution is defined as *the presence of undesirable substances* – the definition favoured by Germany and the European Commission – then no causal connections have to be proven. Pollution becomes a matter of measuring the discharge over time and the concentration of 'undesirable', 'harmful' substances. As should be clear from my arguments in the previous sections of this chapter, this second, seemingly innocuous definition is not without its difficulties either. First there has to be agreement on what constitutes an undesirable/harmful substance and, more problematically, even after agreement is reached on this issue such substances are difficult to pin down: they come from a multitude of static and mobile sources, get transported along many different pathways, are transformed in the process, stay invisible for indefinite periods. Often we do not even know until years later that they are 'undesirable substances'. Both definitions, moreover, assume a technological capacity to quantify and measure pollution and both share a taken-for-granted belief in the possibility of 'green technology'.

On the basis of the arguments I have presented so far in this chapter and in the book as a whole, a truly 'green' technology would have very specific characteristics of which the main one would be that it has to be designed as an integral part of the eco-system cycles of give-and-take, that is, follow the compost principle where 'waste' is simultaneously life source for other beings within the system. This means business needs to be concerned not just with products and production but reproduction, regeneration and, most importantly, the productive forces and principles upon which such re/production is based, in other words, with the *natura naturans* equivalent of economic re/production and ex/change. That is to say, a truly 'green' business knows its product's *Wirkwelt*, its physical bases and the complex temporal cycles of the product's substances. It encompasses in its calculations and concerns not just the complete life cycle of the product and its environment but the life course and the bases of existence of all its substances (see also Hofmeister, 1994 and 1997). Finally, it is concerned with the re-integration of those substances into the environment and seeks a way that *adds value* rather than degrades and depletes.

Clearly, on the basis of these criteria not many businesses and technologies would deserve the green label. But there are enough that successfully pursue eco-design of this kind to glimpse the potential at least for sustainable action. In Britain we have companies such as *Ecover*, headed by a group of young chemists whose products are designed in accordance with those principles. Colborn *et al.* (1996:228) report on *Design Tex*, a US textile firm that began to survey the 7500 chemicals used in the production of fabrics: just 34 of these survived their screening process. Their fabric is now produced in Switzerland in a range of colours and sells at competitive prices. These are just two examples of companies that pursue business and design paths based on ecological principles rather than those of Newtonian science and neo/classical economics. To achieve this they had to shift perspective from space to time–space and take account of the multiple and complex temporalities of technologies, business practices, the eco-sphere and their collective inter-penetrations. Context, networked connectivity and the part–whole relation, re/productive and generative irreversibility, the rhythmic cycles of seasons, life and death, growth and decay as well as indeterminacy, the vast range of time scales and the open future had to become central to their business concerns. That is to say, they had to engage with the timescapes of their socio-natural world of commerce. Given the pervasiveness of time–space dis-tantiation for socio-environmental processes and the concomitant high level of indeterminacy, it is not surprising, moreover, that such businesses think it prudent to take extremely cautious and precautionary approaches to their operations. Finally, and almost by default, their focus had to open out to an expanded time–space, extend the 'rational' pursuit of personal/company gain to considerations of absent others and internalise what had been so radically externalised before. This brings the activities of business much closer than they had been for a long time to the everyday life and social relations outside the time economy of business, to the spheres of life that operate outside the rationale of money and those where money plays only a very auxiliary role.

### Beyond money – 'moonlighting' for the environment

From the basis of a money economy and with the assumptions of neo/classical economics, as we have seen, it is exceedingly difficult to pursue a green economy, technology and business that goes beyond surface paint and rhet-oric. Firms that operate sustainably with reference to ecological principles are still extremely rare. Yet, in the shadows of paid work, commerce and the market economy there flourishes a dark green *oikos* economy. If business associates its newly found commitment to the environment with sunlight and the sunrise then we might call re/productive work that is conducted in the shadow of this charmed circle the moonlight economy of ecological praxis. Unlike Jung, however, I would like this praxis not solely identified with the world of women, even though in quantitative terms the majority of 'experts' in this kind of *oikos* economy are women. The moonlight economy involves not just parenting and caring but voluntary work, environmental activism and single-issue campaigning from animal welfare to concern over road building programmes. What these moonlight practices have in common are, first, that they do not comply with the neo/classical econo-mic model of life, human nature, and society. That is to say, they depart significantly from the individualistic, self-centred, utilitarian, exclusively money-based, short term approaches to life, relationships, socio-economic exchange and re/production. Second, they operate with confidence in the full range of life's temporalities: the timescape is their life world.

The trans-gender activity of gardening can serve as exemplar to show how the everyday pursuits mentioned above activate with ease the green principles of ecology and an *oikos* economy that pose such difficulty at the level of the official money economy, and how they are infused by timescape knowledge and concerns. My analysis is based on a brilliant paper by Heide Inhetveen (1994) in which she uses the principles of gardening as guiding source for sustainable and precautionary action and as a demonstration on how we cope in everyday life with a complexity of time that is not mediated by money. As such her article is particularly successful in making the connection between everyday approaches to time(s) and ecological action, economic practice as *oikos*, and sustainability in terms of reproduction.

Gardening, Inhetveen argues, is about knowing and caring about the future in a time–space distantiated way. Every action in the garden is past- and future-extended and marked by multiple, simultaneous time-horizons: day-to-day care; season-to-season involvement with planting, growing and harvesting; annual concerns about seeds and reproduction more generally; long-term plans and visions about fertility of the soil, as well as the growth and eventual sizes and shapes of plants, their long-term relationships to each other and their mutual interdependence. The time structures associated with gardening are closely tied to the rhythms and temporalities of the seasons, weather, plants and soil and thus bear little resemblance to the invariable, standardised, universal time of clocks and calendars. Gardeners recognise that each plant species has its own time structure and temporal inter-dependencies: for everything there is a right time and this in turn is context-dependent; each aspect resonates with the entirety of the ecosystem, which includes the gardener's actions.

Taking account of these times requires patience, empathy and sensitivity to future needs as well as a holistic way of interacting with and understanding nature's processes. In the garden, order and chaos are always closely related: no work means the garden quickly turns to 'unproductive wilderness'. Gardeners need to have detailed and long-term knowledge of their local conditions and their plants' likes and dislikes in relation to other plants, the soil, the weather, predators, pests, other creatures, possibly even the constellation of the stars. They have to understand the interconnections between their in/actions and the plants, soil, micro-organisms, seasons, weather, the elements more generally and between each of their complex temporalities and temporal needs. Gardeners' actions are thus always integral to the system, the plans and the visions for the future. Gardener, *Wirkwelt* and eventual outcome always form an indivisible whole. With reference to the public economy with its neo/classical assumptions this means notions of objectivity, decontextualisation, externalisation and discounting the future are utterly meaningless here. Neither joy and pride nor responsibility ever get detached from that re/productive unity. Famous gardens as well as family gardens are associated with their creators and a specific place be this ten, fifty or two hundred years after their creation: they are talked about as mum's garden at Pantglas, the garden of Jonathan who wins all the vegetable prizes at the village show, the television garden of the recently deceased Geoff Hamilton who so struggled with his clay soil, or the garden of Lady Eve Balfour who inspired the creation of the Soil Association.

The garden demands identification and empathy before instrumental rationality, sensitivity to context above objectivity, and recognition that each aspect is relevant to the whole just as each action affects the whole. It facilitates a shift in emphasis and priority. Productivity and efficiency take on a different meaning when they are tied to the long-term re/generative and re/productive capacity of the entirety of this eco-system. Key assumptions in the *oikos* economy of the garden are caution, care and precaution, presuppositions that sit uneasily with the rational, calculative, controlling approach of the money economy. Moreover, due to its scale, problems are easily recognised and modified. Even where the prime concern is to grow vegetables to feed the family, the benefits far exceed subsistence: surplus, abundance, joy and the feeling of well-being extend this production beyond mere necessity to encompass pleasure and beauty. The produce may be essential to a family's survival but the time of *oikos* is not or not only equated with money. Instead, time is associated with the rhythmicities of the earth and an acute sense of being part of the re/productive cycle of life and becoming.

Moreover, as the economist Adelheid Biesecker (1994) argues, whether inside or outside the official economy, we are not condemned to competition. In the shadows of the money economy, cooperation reigns supreme. Its success is demonstrated within and between families, in the garden, among environmental lobbyists and activists. In the moonlight economy people relate to each other not as abstract individuals in rational pursuit of their self-interest but as embedded contextual beings with loyalties to families, friends and interest groups, work, society, nature, the earth. Social action, Biesecker argues further, is action tied to morals and responsibilities, duties and expectations. Economic relations of the public kind are not exempt from this. Their 'objectivity' too is inescapably rooted in the realm of ethics. Of course reciprocity, a key assumption of the public economy, is also a feature of the social *oikos* economy but it is complemented there by an acceptance of the needs and rights of those not in a position to reciprocate economically or in principle. Relations guided by the *oikos* economy encompass the needs and rights of children, partners and elderly parents, animals and plants, the environment and the unborn. The right to existence, Biesecker insists, is derivable not from reciprocal action but rooted in life itself. It grants others the right to life and well-being irrespective of their economic prowess.

The issues still to be raised and the connections still to be made are many, but I think the point is made sufficiently for us to appreciate that in the shadow of the money economy – in the family and household, the garden and anti road building campaigns, the peace camps of CND and demonstrations for animal welfare – neo/classical economic assumptions are falsified. Sustainable activity of the deepest shades of green is not just extensively practised, it thrives. Thus, it is here that we need to find our inspirations and resolve, our ideals and standards for ecological business, eco-technology and, if it is possible at all, a dark green market. *Oikos* – the economy of the 'whole house' – is the root meaning of economy. It is this economy of everyday life that holds the secret to sustainable development. It is on the basis of its principles and praxis that we need to be encouraged to moonlight for the environment.

## Time for recap and reflection

The assumptions associated with the linear perspective, Newtonian science, and neo/classical economics, in conjunction with the rationalised time of calendars and clocks, form a powerful, mutually reinforcing conceptual unit. As such, this conceptual conjuncture constitutes the deep structure of the taken-for-granted knowledge associated with the industrial way of life, creating the by now accustomed semblance of certainty and control. It fulfilled this function successfully until environmental hazards started to cumulate and scientists began to be lost for answers. It was hailed as the way forward for all the earth's peoples, until conventional neo/classical economic solutions got exposed as benefiting only an extremely small minority and until it became obvious that politicians lacked not just the means to deal with the ever-increasing environmental problems but, equally importantly, the wherewithal to extend their concerns to time scales appropriate to the hazards created by that way of life. In this changed context of manufactured hazards and uncertainty, the conceptual tools that have arisen with the desire for certainty, mastery and omnipotence now stand in the way of achieving safety and security for contemporaries and successors not just of humans but vast numbers of fellow species. Moreover, as Michael Jacobs (1991:xiii) and numerous others have shown, environmental damage and the manufacture of environmental hazards are not 'an incidental consequence of economic activity', rather, they are a 'central feature of the ways in which production and consumption are currently organised'. The production of hazards feeds and breeds on the assumptions that underpin those activities. To effect relevant change therefore requires reform at the unacknowledged centre of this conceptual system of material production.

Just as physicists at the beginning of this century had to deconstruct their framework of unquestioned assumptions and reconceptualise the very base of their understanding, so economists today are facing the need to begin to consider the unthinkable. In the light of increasing environmental pressure on the integrity of their established theories and practices, the process of reformation has begun. Some economists have started to talk of relations of inequality and power, others of dropping hallowed practices such as discounting the future and externalising all things natural and temporal. Others again have begun to recognise the need for ethics and reflexivity, that is, for acknowledging the role of the responsible actor and her or his implication in the economic scheme of things. The next step will have to be an explicit engagement with the im/material, in/visible *Wirkwelt* of *natura naturans* and an understanding of the economy–environment interpenetration, not only in relation to money and space but also their time–space configuration. There is of course no suggestion to 'turn back the clock', no nostalgia for a pre-money economy. Rather, I am pointing to the pertinence of using the knowledge outside the time economy of money and markets as a source of inspiration for the construction of sustainable futures. I am arguing the need for taking seriously the complex temporal issues entailed in economental processes and their symptoms and for understanding those phenomena in terms of timescapes.

In the remainder of the book I consider the extent to which the habits of mind explored in these first two chapters affect approaches to the environment in policies and practices as they are expressed in political action, scientific and technological innovation and the production of food, as well as media responses to environmental degradation and the production of socio-environmental hazards.

**Plate 3** Manhattan with Brooklyn Bridge by Willi Knaps. *Source*: the photographer.

Part II

# THE EYE OF TIME ON THE INDUSTRIAL WAY OF LIFE

# 3
# SQUARE PEGS INTO ROUND HOLES
## Democracy and the timescapes of environmental politics

## Introduction

When in 1989 Communism collapsed in Eastern Europe, Liberal Democracy became the dominant political persuasion of industrial societies. It is therefore the politics and policies of Liberal Democracy that are seen as ensuring the well-being and rights of citizens while safeguarding the environment against the excesses of technological development and the overzealous economic pursuit of profit. Through their environmental policies Liberal Democracies are claimed to be able to regulate pollution emissions, set 'safe levels' for poisons and contamination, secure the sustainability of the bases of existence – air, water, soil, plants, biodiversity, climate – and to see to it that costs are not externalised, that is, that the true costs to the environment are reflected in the price of goods and services. Liberal Democracy, so it is argued, is the best if not the only system adequate to that task. From a temporal perspective, however, this assumption emerges as naive. The picture gets more complex. Paradoxes replace simplistic certainties: anomalies, gaps, contradictions and incompatibilities abound.

The first difficulty arises from the disjuncture between the popular ideal of democracy and its application in political practice. Here, the history of the concept is instructive. Democracy is a temporal political phenomenon with a long historical tradition. It has its roots in the Greek *demos* which means people and *kratos* which means rule. Raymond Williams (1976) links the early meanings of democracy to a variety of interpretations. These range from 'the rule for and by the people', via 'rule by the many', to a 'generalised opposition to despotic power'. What these interpretations have in common is an association with direct and participatory forms of government. It is this early understanding that persists in the public ideal where democracy means that people have an input into their social destiny, that democracy is ultimately about the rule for and by the people – even if today, due to the size of societies, citizens have to bestow this right to their political representatives. Those entrusted with that right are considered to be answerable and accountable to the people on whose behalf they are enacting the rule for and by the people.

How democracy was perceived throughout history clearly depended on the position of the commentator; it was relative. Since it was the upper and middle classes who had access to historical commentary, it should not surprise us that Williams' (1976:83) research on this political principle showed that until the nineteenth century, democracy had negative connotations: as popular class rule, as power of the masses, it tended to be an unfavourable term. It meant 'ordinary people by force of numbers governed – oppressed – the rich; the whole people acting like a tyrant'. The twentieth century positive meaning of democracy is associated with representative democracy enshrined in the Liberal tradition which secures the right of election and regulates the procedures for choosing representatives. In this modern meaning of democracy, individuals' freedom of speech, assembly and election largely delimit the political role of civil society. The original meaning associated with popular power, and to a large extent also the popular ideal of democracy, as we shall see below, are today associated with a *breach* of democracy.

A second complicating factor emanates from the conflictual constellation of nature, science, technology, and the globalised economy. That is to say, Liberal Democracies face a problem because the quest for environmental 'protection', ameliorative action and sustainability is not easily achieved by a political approach which is so inextricably wedded to the globalised capitalist economy and thus dependent for its legitimation on the rhetoric and achievement of scientific innovation, technological progress and economic growth. This mutuality of science, economics and Liberal Democracy means that a politics focused on the *sources* of environmental degradation and hazards would necessarily entail a fundamental disruption to the *status quo*, a radical shift, that is, in inherent socio-economic assumptions and practices. For Liberal Democracies, therefore, environmental politics constitute a conflict of loyalties of existential proportions.

Third, the proper functioning of Liberal Democratic politics is dependent on operating in bounded social systems within which they have power over their social institutions and their physical bases of existence, that is, in nation states. Its preconditions are similar to those of management which I outlined in the previous chapter. Furthermore, in Liberal Democracies the key decisions and actions, those that affect the populace as a whole, are to be made by elected representatives that can be held accountable. Yet, today, some of the most crucial decisions of long-term effect are not made by our political representatives but by transnational corporations who in turn are not guided by the democratic ideal but by commitment to an unbounded, unfettered global market and the pursuit of short-term profit. As David Korten (1995:99) points out: 'When the market reigns, the corporation is king'. Neither markets nor corporations are elected or socially accountable for their deeds and the impacts these have on society and the environment. The situation is further complicated by the Liberal Democracies' unequivocal commitment to a global economy and the quest for its sustained economic growth, resulting in the tendency to elide the crucial distinction between the rights of money and the rights of people. 'The freedom of the market is the freedom of money', argues Korten (1995:83), 'and when rights are a function of property rather than personhood, only those with property have rights'. Similarly, in a political Liberal Democracy every adult gets a vote whilst in a liberal market rights are tied to money: no money, no voice.

Finally, time is centrally implicated in the problematic interdependence of nature, technology, globalised economy and Liberal Democracy. Time is lived and negotiated in conflict: at home, at work, in the global economy, in the political arena. As such it is subject to a politics of time in its own right.

This conflict, however, is not merely between different scarcities of and needs for time but, as I show in Chapters 1, 2 and 4, between temporalities that operate to different principles, to the variable, rhythmic temporality of nature and the cosmos on one hand and the industrial times of the machine, the laboratory and economic considerations on the other. As such, a multitude of time factors are implicated in environmental degradation and the construction of environmental hazards as they arise in industrial societies from the constellation of nature, technology, the economy and politics. This sphere of contested times, moreover, extends beyond contemporary people and social structures within and outside the traditional jurisdiction of nation states to actions of predecessors as well as citizens and living creatures yet to be born. It thus encompasses beings that have neither influence over nor a means to bestow legitimacy on the Liberal Democracies of the present. These configurations of actors past, present and future, all with concerns, rights and duties, put the classical democratic ideal under considerable pressure. Although not a new issue for governance, the problem becomes accentuated when industrial societies' current actions and those of the recent past foreclose options and livelihoods of untold generations of future beings across the globe. That is to say, the fact that past and present actions predetermine here and now the future presents of such unknown successors demands a new time politic, one more suited to the temporal complexity of contemporary existence. As such, the mutual implications of nature, technology, economics and the democratic ideal constitute a pertinent subject matter for environmental politics. Their collective reconfiguration provides the conceptual and political challenge for the next century. The explicit focus on time gives us a new access to these issues and debates. No longer understood in the singular as the mere context within which political processes take place, time in its complex, multiple expressions – that is, as timescape – can serve as a tool for re-conceptualisation.

In this chapter, I thus use the timescape perspective to illuminate the paradoxes and anomalies in Liberal Democratic assumptions and their expressions in approaches to environmental issues. As in the previous chapters, the detailed academic discussions of democracy and environmental politics are not the focus of my attention here. Instead, I concern myself with political habits of mind and practices associated with this particular contemporary configuration of globalised existence and offer a temporal inflection on the subject. In the first section I briefly consider some temporal features of the Liberal Democracy in the context of environmental concerns. This is followed, first, by illustrative examples of 'Democratic Time Politics in Action' at the national, European and global level and, second, by an analysis of the political role played by transnational corporations and global finance. Crucial to the various aspects of environmental politics are the different ways of relating to and constructing the future. In the final part of this chapter I explore current practices and utopias as they are conducted and constructed in the shadows of nation states and the global economy, visions based on a democratic ideal that can encompass responsibility for the future, on temporal equality and on a common access to the future.

## Liberal democracy as timescape – the political and environmental context

Newtonian science, classical economics and Liberal Democratic politics are closely allied and interdependent. Their shared history and mutually supporting belief systems have elevated that triad to a globally powerful position. As I have shown in the previous chapters with respect to Newtonian science and classical economics, however, the structures and the perspectives they bring to bear on their respective subject matters are beginning to be usurped by the collective outcomes of their actions and assumptions. Environmental concerns about degradation and the industrial production of hazards are one socio-political dimension where these stresses become visible, that is, where dependence on boundedness, focus on individuals, stress on choice and rights, expectation of control, commitment to economic growth and faith in science to deliver certainty and truth – some of the key pillars of Liberal Democracy – are difficult to sustain. In order to avoid repetition I shall merely list in this chapter some of the central issues I have discussed earlier and move quickly to some illustrative examples of the political timescape issues in question.

Let me begin by outlining the changed temporal context of Liberal Democracies that developed at the turn of this century with the standardisation and globalisation of time and with the establishment of the global present. (For detailed accounts of these developments see Adam, 1995a: Chapter 5, and Kern, 1983.) In 1913 clock time became synchronised across the globe. Wireless signals from the Eiffel Tower, travelling at near the speed of light, displaced variable local times and imposed instead one uniform, hegemonic world time for all. Thirty years earlier Greenwich was installed as the 0 Meridian and the earth divided into 24 equal time zones, each one hour apart. This standard time too brought to an end the myriad of contextual times and dates used by the peoples of the world. That is to say, both these developments rationalised time and severed it from local contexts and conditions: one hour became one hour irrespective of season, time of day and place. Both facilitated global synchronisation as well as certainty and predictability in social interactions. In conjunction with the world-wide net of wireless and electronic communication which brought about the global present these innovations in social time smoothed the path to global integration, to the contemporary global network of political, economic and socio-cultural activity. Moreover, this package was vigorously exported to cultures who had very different relations to time. It was imposed irrespective of whether or not these temporal innovations were welcomed or rejected there: non-compliance spelt automatic exclusion; it meant being constructed as 'other' and therefore in need of 'development'. The globalised rationalisation of time and the creation of global simultaneity can thus be seen as integral components of the wider global colonisation strategies of this century, such as the global market and global finance or the invention of poverty and its associated quest for 'development' facilitated by such global institutions as the World Bank. World time, standard time and the globalised present consequently constitute the temporal context for contemporary global diplomacy and trans/national politics.

The political structure of Liberal Democracies, however, is a left-over from a different historical era. It is not designed to function as a global institution. That is to say, Liberal Democracies are tied to the politics of nation states. One of their principle assumptions is boundedness – of territory and of period of election. Thus, although such national governments function within a global context of time, communication, finance and markets, their sphere of action is delimited within a particular time and space. Thus, when environmental hazards, created within the jurisdictional time–space of a particular Liberal Democracy, transcend those time–space boundaries, the impact is in effect externalised: to other nations and/or to successor generations. As is the case with economic externalisation (see Chapter 2), the problem is shunted along, moved outside the sphere of responsibility. Without a higher time–space authority, hazards externalised in time and space are no longer recognised in principle as the concern of the offending nation's representative government in office. The clearly necessary higher authority is to be provided by such transnational institutions as the European Union and the United Nations. Much of their authority, however, is related to space. In other words, we have as yet no socio-political body that safeguards time and most specifically the future. Given that most problems that are being created now will not be recognised as dangers and thus will not emerge as hazards for an unspecified period of time, the boundedness of responsibility enshrined in Liberal Democracies is highly problematic for engaging with the environmental issues addressed in this treatise. Equally questionable, of course, is the assumption that time–space distantiated hazards can be 'managed' according to traditional principles. I have detailed this particular difficulty in the previous chapter.

In addition to boundedness in time and space, Liberal Democracies are used to operating in quantifiable contexts. Their territories, people, institutions, traffic, crime, budgets and GNP's – all can be counted and measured. They are quantifiable. However, this basic principle too is rendered largely inoperable with environmental hazards. The quest for quantification of the problem becomes futile, as I have indicated in previous chapters, when the processes involved are time–space distantiated, non-proportional, marked by contingency, time-lags and periods of invisibility, or when they are so drawn out that their beginnings and ends can no longer be held together, neither in theory nor in practice – as was the case, for example, when governments began to plan their nuclear futures (see discussion pp. 111–13; Welsh 1994). In the face of such characteristics, not even the most competent government scientists can calculate levels of risk and/or safety, as has been amply demonstrated by the BSE crisis (Chapter 5), in the aftermath of the nuclear explosion at Chernobyl (Chapter 6) and by the release of genetically modified organisms (Chapter 7). Irrespective of the difficulty, however, the tradition of quantification is taken for granted. Quantification is the basis to most environmental regulations, guidelines, taxation, and legal definitions. Without quantification a key pillar is taken out of the political sphere of competence, thus endangering governments' capacity to plan, act for, and manage the environment. Although globalisation of clock time rationalised time to the point of its being the quintessence of quantification, therefore, the globalised time of environmental hazards eludes measurement and quantitative assessment: networked connectivity, differential uptake in land-based and aquatic creatures and plants, airborne dispersal, time-lags and periods of invisibility, all create the opposite effect. Similarly, while the clock-time context of Liberal Democracies is globally constituted through de-contextualised world time, standard time and the global present, their time–space sphere of operation is limited by their territory and the period for which governments are elected. These are just some of the time-based anomalies and paradoxes that abound in the contemporary socio-political sphere of environmental politics in Liberal Democracies. Others relate more explicitly to the construction of the future.

Today, open-ended futures are being constructed and foreclosed with the conceptual and political tools of bygone centuries without a means to take account of and be answerable to citizens now or in the future who live outside the limited jurisdiction of governments' terms of office. That is to say, the future-creating activities of Liberal Democracies have to be seen in the context of governments with a mandate to act on society's behalf that extends to a mere four to five years. This means a time-span of responsibility that does not even cover the building phase of a nuclear power station, let alone its period of decommissioning and even less the time-span of its radio-activity. The effects of decisions often outlast the governments who made them by many generations. Moreover, since elected representatives are responsible to their electorate and since it is this electorate who bestows legitimacy on a government, the rights of future and distant people who cannot enact that (power) relation are 'discounted' in a way that is analogous to the discounting of the future in economic processes. The effect seems the same: the further away the potential hazards or degradations, the less they count for considerations in the present.

Finally, Liberal Democracy is a political system whose promises to the electorate are frozen in time, tied to the time of the election and bounded by the time-frame of election cycles. Whilst contexts and political problems during a term of office are contingent and subject to change, manifestos and election promises are fixed in time and space. In practice, of course, the promises get adapted and the periods of political influence necessarily extend far beyond the term of office, beyond the parliamentary present of four to five years, to the open-ended realms of both past and future. Temporal transgression of the term of office is in fact inevitable since, as I illustrate below, no matter how radical the shift in government, the voices and visions of the past are enshrined in the future-constructing present of the body politic through dependence on the global economy and/or a permanent civil service and/or incremental changes in laws and policies.

These temporal issues are fittingly exemplified at the national level by some of the recently released papers of the Powell Committee on Britain's nuclear policy, discussed by Ian Welsh (1994), at the international level by the European Union's environmental politics and at the global level by the United Nations' Conference on the Environment and Development (UNCED).

## Democratic time politics in action: environment and technology in national and transnational policy

### *The Powell Committee*

The Powell Committee was set up in 1962 by the British government to review the future of nuclear energy in the UK. The papers of the proceedings of that committee which had been classified under the Official Secrets Act have been released recently after their obligatory thirty year period. Under the chairmanship of Sir Richard Powell, the committee whose 'membership was confined to representatives from the Treasury, Ministry of Science, the Board of Trade, Ministry of Power, Atomic Energy Authority, the Central Electricity Generating Board and the South of Scotland Electricity Generating Board' (Welsh, 1994:45) was to select the type of nuclear reactor and determine the size of the programme for Britain's nuclear future. Its brief was to arrive at costings up to and beyond the year 1985 and to take account in their decisions of 'the interests of the economy as a whole' (Welsh, 1994:46) as well as Britain's position in the global market place. Whilst the operating costs of the reactors under consideration were calculated on the basis of an assumed thirty year life-span, the time-scale of decommissioning and associated costs did not figure on the agenda.

The UK approach to nuclear policy emerging from these released papers raises a number of issues relevant to this analysis. With respect to the enactment of democratic principles, these relate to the secrecy, the composition of the membership, the role of the treasury and economic expediency, 'big science' interests, and the time-scale of influence. Of particular note, with respect to time, are the committee's relationship to the future, its calculations and assumptions, the cumulative rational of its decisions, the long lead-times of nuclear power in relation to the speed of technological development, and the operating time-spans and costs in relation to the de-commissioning time-spans and costs. I very briefly consider these in turn.

The Powell Committee was established as a Cabinet Committee, which meant on the one hand that it had direct links to Whitehall and on the other that its existence was so secret that Members of Parliament only heard about it by rumour and through personal contacts (Welsh, 1994:44). MPs, therefore, had no formal input to that committee. Its secrecy in terms of its setting up, membership, terms of reference, proceedings and deliberations precluded both democratic procedures and any hope of resonance with the popularly held democratic ideal. Moreover, whilst it is quite clear from the papers that Britain's economic interests and its standing as a leading scientific nation had to be enhanced by any proposals the committee was to put forward, considerations concerning the well-being and safety of citizen 'end-users' were significantly absent from the deliberations. The democratic credentials of the committee, therefore, cannot be established on this basis either.

Finally, the temporal issues surrounding the work of that committee substantiate further the undemocratic nature of Britain's nuclear policy as it arises from these papers. The committee was charged to come up with recommendations covering a time-span of twenty years and beyond. This has to be appreciated against the political backdrop of a Liberal Democracy that offers elections to its citizens every four to five years. Furthermore, the *actual* influence of the deliberations of the Powell Committee, of course, affected not just the UK but a global community of citizens to varying degrees and for an open-ended, long-term future. It thus foreclosed options for people yet to be born, for future generations without means of redress. This meant, the committee's work prevented due political process on this matter nationally and globally not just for decades but hundreds, possibly even thousands of years hence: radiation from leakages and accidents found and will find its way into the food chain and the genetic make-up of future generations. Within its own national boundaries, furthermore, Britain's nuclear policy had, has and will have knock-on effects on the policies of future governments. At the very least, nuclear power plants have to be de-commissioned. Spent fuel has to be reprocessed and the waste 'secured' for an open-ended future.

Nuclear power is an inevitably long-term and time–space distantiated socio-political and economental affair. With their long lead-in time, nuclear reactors and other installations invariably are outdated before their completion, with the result that governments tend to be tied in their commitments to antiquated technologies and 'dinosaur projects'.

> Projects such as reactor programmes ossify a set of prevailing scientific technical and socio-political conditions embodying them in

concrete and steel and enmeshing society in the web of associated techniques required to deliver their technical potential.

(Welsh, 1994:51)

Moreover, due to the combination of the high level of investment and the incursion of an accumulation of past decisions into present considerations and judgements it is extremely difficult for governments to extricate themselves from their predecessors' policy decisions, hence the current situation where the dramatically altered context has not resulted in corresponding policies. In the face of deep public mistrust about the safety of this source of power, a radically altered economic context, and ever accelerating changes in scientific knowledge and technological innovation, the attitudes and assumptions displayed by the Powell Committee may have given way to a more tempered political approach, but nuclear power is still with us, not abandoned but merely placed on the back-burner, in readiness for reactivation in times of future socio-political need.

Some of the issues and implications for democracy are repeated at the inter- and trans-national levels of the European Union and the United Nations Conference on Environment and Development. To avoid repetition, I will only raise points that have not already emerged from the discussion on UK policy formation in relation to nuclear power.

### The European union: environmental politics – dilemmas of democracy

Since environmental hazards do not respect the political boundaries of nation states and the time-frames of election cycles it has become necessary to find ways of engaging with environmental issues that transcend national politics. The European Union (EU) is one political body where transboundary policy is formulated and enacted. Its role with respect to the environment is both active and reactive. It considers itself a driving force in the safe-guarding of a healthy environment and an arbitrator for the complex negotiations of compromises and harmonisations between member states. In this dual capacity, by 1994, the EU had passed in excess of 200 laws governing process and production norms in relation to water, air and waste (Taschner, 1994:2). That is to say, with reference to the environment, the European Commission seeks to take a pro-active role, but this tends to get diluted in the actual negotiations with member states whose representatives still hold the decision-making power, a privilege that is exercised behind closed doors and in a context of a lack of democratic accountability.

Analysts of the environmental politics of the EU tend to agree that its approach to the environment is marked by a 'democratic deficit' (Baker, 1996; Bongaerts, 1994; Florenz, 1994; Hey, 1994; Taschner, 1994; Telkämper, 1994), yet they differ in the degree of pessimism and the emphases they put on particular causes of that state of affairs. An obvious candidate is the structural arrangement of EU policy making. Thus, as Susan Baker (1996: 10–11) points out, the EU has an executive and a Parliament but lacks the democratic means of linking the two: the Council of Ministers and the European Council make policies for which they are accountable in a limited sense to the elected European Parliament but effectively to no one. Christian Hey (1994:8) calls the European Parliament the democratic fig-leaf of the EU since it has only a limited veto and control function but no active input into policy making. Those structural arrangements, however, seem to be only one of a range of sources for this widely observed democratic deficit.

The most deeply rooted and troublesome cause for this diminished political accountability which permeates environmental policy from global to national level is the unswerving commitment to a global market and the economic growth imperative. (See Chapter 2 for a discussion on the complex relation between a commitment to economic growth and concerns over environmental degradation and hazards.) At the level of EU policy, the *common market* has been and will continue to be the primary consideration of the EU. Environmental policy, therefore, can be created only in the context of that 'higher authority'. That is to say, environmental policies are only acceptable to the extent that they do not interfere with the economic goals of the Union. Where environmental policies potentially interfere, as in the target of a 40 per cent reduction in $CO_2$ emissions, a healthy antidote of transport policies that will vastly increase the network of motorways and double air traffic by the year 2005, for example, will swiftly restore the balance in favour of the market (see Gebers, 1994:33; Hey, 1994:10; Taschner, 1994:7; Telkämper, 1994:14–15). Moreover, it was a 1984 EU Directive that introduced the idea of 'best available technology not entailing excessive cost' for a measure that was to ensure air quality and emission control standards at minimum cost. Thus, we need to recognise, radical change that would get to the economic and technological root of the environmental crisis is *not* on the EU agenda; measures that increase the economic potential, in contrast, are – hence the keen promotion of 'green' and 'clean' technology as market stimulants: 'what is good for the environment can also be good for business' (Industrial Competition and Environmental Protection, SEK, 92, 1986, quoted in Bongaerts, 1994:34; see also Chapter 2 as well as Adam and Kütting, 1995). The growth imperative of the market, global finance, and 'big science' technological developments, however, do not only play a central role in the responses to environmental crises but they are also responsible for them in the first place and crucially implicated in their discordant temporalities. Moreover, none of these social forces, as I have already suggested with reference to the Powell Committee, are subject to democratic influence. All operate outside the time–space of due political process.

The EU prides itself in working with long time frames: it thinks and plans environmental provision in decades. The economic system and the technology it supports, however, create futures for centuries and even millennia. The EU sees itself spearheading future-oriented policy. In practice, however, much of the concern is about damage limitation in relation to the unintended consequences of past actions: past and future have become porous and set the agenda for the present. Options are increasingly foreclosed which severely limits the EU's spearheading potential (on this topic see also Nowotny, 1989/94: Chapter 2). In addition, studies show that only environmental issues that fit into already existing schemes have a chance of getting on the EU's environmental agenda (see Baker, 1996). Moreover, as is the case with the UK's nuclear policy, the EU's style of policy making is incremental. This means, argues Susan Baker (1996:17), that highly visible, controversial projects with a potential for unsettling the *status quo* have a low success rate. The time-horizon of decades and the incremental, thus serial, linear and past-orientated logic of EU environmental policy has to be seen in the context of the complex temporalities of environmental problems discussed in this book: time frames of environmental processes that range from nano-seconds to millennia in conjunction with networked connectedness which makes the processes non/linear, non/proportional and time–space distantiated, which means, input and output are not necessarily connected by time and space. Moreover, environmental hazards are marked by multiple time-lags: between source – symptom – recognition – acceptance as problem – policies – actions – impact of actions – arrest and/or potential recuperation.

Given these glaring discrepancies between time politics and the environmental problems they address, it seems obvious that not the environment but the economy and socio-political expediency are the driving forces behind the environmental agenda of this transnational political body whose purpose of existence is to be a key player in the global market. Placing environmental problems on a global agenda, as we shall see below, does not overcome but magnify the limitations encountered with reference to EU environmental politics and its associated democratic deficit. Economic considerations are paramount. The global market rules supreme. The demarcation between Liberal Democratic politics and the logic of the global market has been irreparably eroded.

### United Nations Conference on Environment and Development – Rio 1992

The 'Earth Summit' at Rio – a logical progression from the work started by the Brundtland Commission in 1987 – had as part of its agenda the restructuring of consumption patterns, economic priorities, and political structures. It even incorporated peoples yet to be born into the spheres of present-day local, national and global responsibility. UNCED produced five separate agreements of which two were binding (but not agreed upon by all the States) and three provided visions for the future which, on the one hand, carried a certain amount of moral pressure on governments across the globe to take heed and, on the other, gave strength to local, non-governmental actions. The Earth Summit itself, however, demonstrated the depth of divisions between nations and their ruthless pursuit of national advantage. Change in patterns of consumption, eradication of poverty, the world-wide implementation of the Precautionary Principle, and the cessation of the transfer of damaging substances to other states – all enshrined in the Rio Declaration on Environment and Development (see Quarrie (ed.) 1992:11–12) – were each considered to be beyond the boundaries of nation-based negotiation. Everyone, it seems, set out to secure terms that best suited their national interests and sought to achieve them without disturbance to their *status quo*. For the powerful industrial nations this meant leaving untouched the global system of trade, finance, and debt collection. For those at the receiving end of the policy of development, globally instituted negative equity and net resource flows to the North, the issues raised at Rio merely offered some leverage to secure a more equitable distribution of resources. Thus, the Earth Summit was marked by a gulf between the vision put forward by UNCED – enshrined in the Rio Declaration's 27 principles and Agenda 21, the UN's detailed action plan to operationalise sustainable development – and the willingness of States to negotiate over the implementation of the necessary changes.

The incredulity and rage felt by participants from Third World countries about the rich nations' hypocrisy and imperialist attitudes is pertinently expressed in a paper by Vandana Shiva (1993) in which she puts forward a widely agreed upon argument against environmental management on a global scale. To elevate locally caused environmental problems to global and future status, she suggests, allows us to lose sight of the *causes* of environmental degradation and conveniently shifts the burden of responsibility from the perpetrators onto the shoulders of Third World countries. Thus, worries about the masses in the underdeveloped world wanting fridges, for example, diverts attention from the wealth-creating manufacture of fridges which are to be sold to these newly emerging markets and the associated ozone-depleting chlorofluorocarbons (CFCs) or their highly problematic replacements. Shiva and other like-minded environmental commentators from the South suggest further that globalisation of environmental problems brings with it a damaging reduction in sovereignty and local control for those in powerless socio-political positions.

> The 'global' in the dominant discourse is the political space in which a particular dominant local seeks global control, and frees itself of local, national and international restraints. The global does not represent the universal human interest, it represents a particular local and parochial interest which has been globalised through the scope of its reach. The seven most powerful countries, the G-7, dictate

global affairs, but the interests that guide them remain narrow, local and parochial.

(Shiva, 1993:149–50)

She expresses the widely held opinion of those at the receiving end of development and aid when she concludes: ' "Global" concerns thus create the moral base for green imperialism' (Shiva, 1993:152). The World Bank, the most powerful and least accountable institution on this globe, is entrusted with the protection of the global environment (through the GEF, Global Environmental Facility, one of its subsidiary institutions). Yet, to the peoples of the South the World Bank is one of the key players in the systematic destruction of local habitats and livelihoods.

Shiva (1993:154) thus speaks for many when she asks for the democratisation of 'global' interests and institutions and a strengthening of local rights. 'What we need to ensure', she writes, 'is that no World Bank decision affecting tribal resources is taken without their prior informed consent' (Shiva, 1993:155). This, of course, would require the sort of changes that were so successfully resisted by the powerful few at Rio, are non-negotiable for the EU, and were explicitly counterpoised by the recommendations of the Powell Committee. Moreover, the issues raised by Shiva and many of the environmental activists of the South are of concern to all of us. They are, as I argue below, truly *global* matters of concern. I am referring here to the role in national politics of transnational corporations and trade agreements as well as global finance and institutions such as the World Bank.

## Global governance by corporate values and money

Liberalism is an ideology and approach to life that comes in many guises and permeates both economic and political relations. In the economic sphere it is variably called economic liberalism, neo-liberalism, market liberalism or corporate liberalism. Economic liberalism shares with Liberal Democracy an emphasis on individual rights, freedoms and choice: as political agents individuals are considered free to vote, which gives them the right to choose their government and the policies they want pursued on their behalf. As consumers they are argued to have the freedom and the right to choose between products. Through their preferences individuals exercise control and influence the direction of politics and the economy. Through their choices, moreover, these 'free' individuals are considered the makers of their own destinies. In countries such as the USA and the UK the alliance between economic and political liberalism is particularly strong, melding the two social spheres to a point where there is cause for concern. It is these overly strong ties I want to focus on here as they have a deep significance for the diminishing political credibility in these two countries and for the ever-increasing gap between political practice and the democratic ideal. (As I am writing within a UK context my comments are made with reference to this country unless explicitly stated otherwise.)

In countries such as the UK and USA where there is such a strong political commitment to the 'free' market, there is a tendency for the economic valorisation of growth, privatisation, globalisation, deregulation and freedom of (business and consumer) choice to be almost equated with political rights. This has the effect that much of political power is actually abdicated to the market on ideological grounds. Where sovereignty is otherwise a primary concern it seems happily sacrificed on the altar of international free trade. In a brilliant book on the political role of corporations, David Korten details this rationale in the following way:

> Interwoven into the political discourse about free markets and free trade is a persistent message: the advance of free markets is the advance of democracy. . . . The logic is simple: in the free market people express their sovereignty directly by how they vote with their consumer dollars. What they are willing to buy with their own money is ultimately a better indicator of what they value than the ballot, and therefore the market is the most effective and democratic way to define public interest.
>
> (Korten, 1995:66)

As I already indicated in the introduction, however, the difference between the political and economic way of exercising choices and rights is significant: while political rights are expressed by the vote, economic ones depend on money. In the world of economic exchange, that is, choices and rights evaporate when a person has no money. This means markets are inescapably biased in favour of wealth and, importantly, more money means more sociopolitical power which makes the largest transnational corporations, banks and financial institutions extremely powerful. Measured in money terms, for example, the power of national governments is dwarfed by the amount of money that flows through corporations and financial institutions. In the region of $800 billion get traded each day in the international currency market alone. That is enough money to feed the entire world for two years! Beyond this vast discrepancy in money power there are a number of other dimensions including some pertinent temporal features that all work together towards the creation of an ever increasing democratic deficit.

We are dealing here with an ideological confederation of institutions and practices that comprises transnational corporations and international trade agreements, globalised financial markets and global institutions such as the World Bank and the World Trade Organisation. Despite their powerful roles in our lives, however, these institutions are neither accountable nor liable for their socio-environmental effects. Far from being democratic in the political sense of social accountability and responsibility, they are answerable only to the authority of money. In the economic pursuit of unimpeded flow and accumulation of money, they function on the basis of rationalised, de-contextualised socio-environmental *irresponsibility*. It means, this ideological confederation is marked by a common commitment to the creation of money and *an explicit non-allegiance to people and places*. Money is both the lifeline and the exclusive measure of value. The potential for maximum profit dictates where corporations, for example, place their operations, move their finances, deposit their pollution. It further determines the proportion of people hired to staff fired, of expansion to 'downsizing'. Where money is the goal, means, outcome and final arbitrator between values, people's needs as well as concern for social well-being become a source of economic inefficiency and weakness: thousands, even millions of people are being cast aside by corporations that no longer have a need for them because the job can be done more cheaply by a machine or by workers in another part of the world. Equally, when environmental regulations stand in the way of maximum profit it tends to be cheaper to up the operation and move it to a country that has less stringent environmental protection laws than to comply with the regulations. Quite clearly, the larger the corporation the more leverage it has to bargain with governments and to press for deals favourable to its pursuit of maximising profits.

Environmental commentators have designated this a 'race to the bottom', a downward spiral where the financially optimal and socio-environmentally worst condition becomes the baseline for economic relations and modes of production: lowest wages, lowest safety standards, lowest company expend-iture, lowest environmental protection and concern, elimination of barriers and regulations. From this perspective, Margaret Arwood (1993:92) writes, 'I believe the free-trade issue has the potential to fragment and destroy the country in a way that nothing else has succeeded in doing' (see also Brown, 1993; Korten, 1995; Mander, 1993; Nader, 1993; Wallach, 1993). Economic commentators, as one would expect, take the opposite view, extolling the virtues of free trade. A quote from an essay by Frances Cairncross (1995), a journalist with *The Economist,* is instructive here.

> But trade barriers do not help industry to compete in world markets. And they do nothing for consumers; indeed, by keeping out low-cost goods, trade barriers tend to leave people worse off than they would be otherwise.
>
> (p. 233)

> Allow one country to discriminate against one production process on environmental grounds and the result is a whale-sized loophole. Where would discrimination stop?
>
> (p. 236)

Clearly no government should be allowed to apply general trade sanctions unilaterally to bully another to adopt a particular environmental policy.

(p. 237)

While it has to be noted that the deterministic gloom of environmental commentators is not borne out by vast numbers of examples world-wide that demonstrate people's capacity to act in defiance of these economic downward pressures (as illustrated throughout this book), Frances Cairncross' view is illuminating for its expression of the blinkered economic perspective on socio-environmental issues. Focus on some of the time–space issues involved can shed some additional light on this intractable situation of incompatible perspectives. The International Free Trade Agreement, transnational corporations and global finance are chosen as exemplars for the wider points.

The very essence of the globally constituted Free Trade Agreement, policed by the World Trade Organisation since 1995, is its dissociation from time and space. In tune with the globalisation of time – that is, world time, standard time and the global present – locality, context, seasonality and history are rendered irrelevant. *Absolute decontextualisation is the ideal condition for money to flow freely and for capital and operations to be moved unencumbered where the circumstances for wealth generation are optimal.* In such decontextualised conditions, real people living in particular places with specific needs are sidelined out of the frame of reference: they have no place in a decontextualised world. In addition, as I have argued in Chapter 1, without context and boundary conditions we fall under the illusion that time is reversible, a belief that facilitates risk-taking behaviour on the assumption that mistakes can be undone and damaging practices reversed. With each decontextualising move, we distance ourselves further from the operations of ecological processes for which the uniqueness of context and interactions is inescapable: the time of year and day, location, as well as all of an ecosystem's specific interactive participating organic and inorganic 'components'. Absolutely nothing is the same for anything or anybody in the living world and this includes human societies. Everything is contextually unique. Only a political system reared on Newtonian physics, rooted in the linear perspective and steeped in the assumptions of classical economics could try to persuade us otherwise. But then, who is convinced beyond those who stand to gain and improve their wealth? As Cairncross (1995:226) admits, 'it has never been easy to persuade ordinary voters that freer trade is a cause worth defending. Its economic benefits are readily ignored when foreign competition threatens local jobs'.

A politics that steamrollers such ideologies through against the wishes of civil society and against even its own habit of vigorously defending sovereignty is therefore deeply disconcerting. There is cause for concern when the power to make laws within and between sovereign states and on behalf of their citizens is ceded to the interests of transnational corporations, science, and economic principles. Where a country has fallen foul of the International Trade Agreement, we need to appreciate, science is the only acceptable arbitrator for situations where a conflict of interest arises between money and the well-being of people and/or the environment. 'The offending country', explains Korten (1995: 174), 'must prove that a purely scientific justification exists for its action. The fact that its citizens simply do not want to be exposed to higher levels of risk accepted by lower World Trade Organisation (WTO) standards is not acceptable to the WTO as a valid justification.' Since, however, traditional scientific proof is unobtainable under conditions of time-distantiation, invisibility and time-lags, as I argued in the previous chapters, recourse to science is a neat and effective way of rendering the law-making capacity of sovereign states inoperative (see also Chapter 5 with reference to the UK BSE crisis).

With respect to transnational corporations, which are of course equally committed to decontextualisation, I want to highlight some additional time–space features. Prominent among these is the discrepancy between their approaches to time and space respectively: while their space of operation is the globe, their time horizon of concern is exceedingly narrow. Short-term profit is prioritised over long-term gains. The present takes precedence over the past and future. The future is discounted. A second characteristic relates to the disconcerting fact that corporate culture is renown for its abstraction from people-based history. Its history is written in money, takeovers and growth patterns, and thus the transnational corporation has no commitment to past, present or future employees, no loyalty to its country of origin or the societies within which it operates: money, not people, defines efficiency, expediency, and appropriate action. This opens up another disjuncture, this time between the pursuit of short-term profits and the long-term effects of corporate economic activity on local cultures and the environment. For local communities the (often abrupt) disappearance of a major employer to more profitable locations creates social havoc and devastation, engendering a volatile mixture of anomie and alienation. As living rather than abstract entities, people build up loyalties and commitment, nurture relationships and special interests, develop specialist knowledge – all slow and long-term processes. When these long-term interdependencies are suddenly dissolved, social structures evaporate, leaving members of communities without the support and stability essential to communal life. The time for recovery tends to be proportional to the devastation. Third, members of living communities and families are not exchangeable, whereas the members/employees of corporations are. For the corporations they are like parts of a machine, which means their function can be achieved by another human part. The stability of the structure is maintained as long as sufficient money flows through the corporate veins and arteries – shareholders and the free market, so the assumption goes, sustain the corporations' source of life.

With global finance these time–space features are amplified: abstraction and decontextualisation are total. Financial trading occurs non-stop, globally, at high speeds – a sleepless, restless computer-based affair. Investment results are published daily; a month is an eternity in financial market terms. The biggest money is to be made in exchange rates speculations where money gets moved around at extraordinary speeds, often not remaining in one place for more than a few seconds. Fortunes are made over a time-span of a few seconds in an electronic context of simultaneity and instantaneity. The short time-spans of the deals associated with global money markets are legendary: time–space and money seem to stand in an inverse relation. Speed, indeterminacy, global connectivity, reflexivity of the processes, and the vast sums of money involved create a volatile condition for trade. Financial markets thrive on it: volatility is their source for opportunity and profit.

This particular temporal mix has to be seen in the context of Liberal Democratic governments who seek stability within the boundary of their four to five year term of office as a precondition to successful economic and socio-legal planning. Such governments operate in the context of socio-environmental continuity but with their mandate temporally delimited, which means that their activities, by definition fragmented, are focused on time-horizons that bear no temporal relation to the issues in question. On the basis of this structure, therefore, political emphasis tends to be biased towards short-term fragments of a-historical stretches, disembedded from the continuity and past–future extension of community and ecologically embedded life. Framework upon framework upon framework of time–space are superimposed upon each other, each with its own variety of temporal extensions and tempi, its own timing priorities, its own speeds of operation and change. All are mutually implicating and influencing in a dynamic pattern of interrelation. In their mutual incompatibility they are lived and negotiated in conflict, that is, they co-exist in friction and a hierarchy of status. As such, the timescape of Liberal Democracies is subject to inequalities and power differentials which governments ought to regulate instead of abdicating to the higher authority of science and money. This lack of an explicit time politics, therefore, is crucially implicated in the gloomy prognoses of environmental commentators and in the loss of legitimacy experienced by Liberal Democracies in the light of increasing environmental hazards and degradation. Recognition of the timescape of the contemporary Liberal-Democratic, scientific and economic configuration on the one hand, and the environmental degradation and hazard creation on the other brings into sharp relief the need for new structures of governance.

## Temporalised democracy in action and concern for the future

The Rio Declaration encompasses some powerful ideals and Agenda 21 some pertinent suggestions for change. Despite the extensive resistance at national level, these ideals are being widely instituted at the local level. The EU is producing environmental directives and regulations that not only run counter to its purpose of existence but are also clearly in breach of the Free Trade Agreement. At a recent conference, Greenpeace and Friends of the Earth were talking to and debating with transnational corporations. The chairman of Tarmac, a transnational corporation with a strong portfolio in road building, sided with the arguments of the road protesters against the strategies of the UK government (Conference, *Reporting the Environment*, Cardiff, 19–21 May 1996). Even the newspapers, overall the most conservative force in this particular network of social actors, are not just reporting on environmental activism, road protests and animal rights campaigns, they go further and inform citizens of their rights as well as instruct them in the rudiments of environmental protest. Are we therefore not already experiencing changes in political culture in the direction of some of those goals? Could it not be that those cogent ideals have rippled their way into governmental institutions and that we are seeing the first signs of attempts at their implementation in such examples of democratising developments as the EU's principle of Subsidiarity, Bavaria's Vote on 1 October 1995 over increased democracy and participation at local government level?

Liberal Democracy is not just the contemporary, hegemonic democratic political practice; it is also, as I suggested in the introduction, a complex popular *ideal* that has the capacity to mutate with the context in which it is applied: interpersonal, organisational, political. The core of this ideal has retained its close ties to the original Greek meaning that it is (or should be) citizens who determine the orientation of public life. As such, the ideal is associated most specifically with the right to be heard, to have a voice that can influence the powers to be and affect their decision-making processes. It encompasses the safeguarding of individual and collective rights, most importantly the capacity and right to be able to intervene when those entrusted with government are acting irresponsibly or against the national/ global interest. This, in turn, entails trust that those who have been given the mandate to act on behalf of the populace actually execute the popular will and that they are accountable for their actions and particularly their failures.

Liberal Democracies that see the world through the eyes of economics, judge political success on the basis of money and conceive of their task in fragments of four to five years of governance, as I have argued, have difficulty in representing the needs of people, the environment and future generations. For these tasks, politicians will need to defer not to economics and science but to the experts in long-term vision: parents and grandparents, gardeners, foresters, indigenous cultures, farmers who have retained a close relationship with the land and preserved a sense of continuity, young people committed to the welfare of animals. Environmental groups, on this count, would not automatically count as experts in this field. Their credentials would have to be established on a case-by-case basis. In the UK, for example, environmentalists have set up a *Forum for the Future* which has the 'explicit purpose of taking a positive, solutions-oriented approach to today's environmental problems' (Jacobs, 1996:137). Until its members engage with the temporal issues, however, I am not convinced of the alliance's promise as a 'forum for the future'. In Germany, the Federal Government has established a *Futures Commission (Zukunfts Kommission)*. Not surprisingly, it seems to be exclusively concerned with the creation and maintenance of jobs. While this is clearly a future need, its potential for success is extremely low as long as the problems are tackled from an economic perspective and without a timescape approach to the past, present and future of jobs. What then is the likelihood of such a radical change in perspective and politics? What are the chances of Liberal Democracies taking a temporal perspective and recognising their need for people expertise in an area in which they are particularly ill equipped to lead from the front?

I know such a radical change is difficult to imagine but then, in our very recent history in Europe we have experienced popular democracy. In 1989, with the collapse of East European communist regimes, impossibilities became reality. Citizens enforced the democratic ideal: to determine the orientation of public life, the right to be heard, affect the decision-making process, and intervene when those entrusted with government are no longer acting in civil societies' interests. In the light of these encouraging developments and the increasing realisation of 'miracles' we need to embrace the conviction that *anything is possible* as long as enough people want it and organise their lives accordingly. It does not mean everybody has to turn into a particular kind of eco-warrior – far from it. Diversity is strength and every tiniest action irreversibly alters the whole. It does mean, however, that we need to make the taken-for-granted visible, recognise the hidden workings of the science–economics–Liberal Democracy configuration and appreciate its drag and pull-down effect on environmental action and concern for the future.

## Time for recapitulation and reflection

Focus on the temporal complexity of the time politics of Liberal Democracy in the context of a commitment to the industrial way of life has brought to the surface a number of pertinent issues: it has pointed to the multiplicity of times that need to be taken account of at any one moment of decision making and in any one policy solution. It has shown the complex past–future penetration of the present which inescapably transcends the four to five year period of governments in office. It has illuminated the complex configurations of peoples and beings past, present and future which are implicated in and affected by the decisions and policies of Liberal Democracies today. Finally, it has explicated the deep conflict of temporal principles between society and nature on the one hand and the imperative of economic growth and science on the other. It has shown the implications of that discord working through the time politics of Liberal Democracies. This suggests that the spatially oriented politics of nation states with their territories and borders, their spatial political alignments of left and right, and their linear (spatial) conception of time are ill-equipped to encompass and respond to the temporal complexities arising from globalised environmental hazards created by the industrial way of life. The answer, I am afraid, is *not* to retreat to the local in a response to the imperialising global reach of a few powerful nations and in solidarity with the pertinent analysis produced by environmentalists of the less powerful countries at the receiving sharp end of those problems. If the industrially produced hazards are global then 'thinking locally and acting locally' (Esteva and Prakash, 1994) – the current adaptation of the slogan 'think globally, act locally' – cannot be sufficient. Instead of falling back on the familiar logic of binary thinking and either–or options, I suggest we raise our temporal awareness and scrutinise social and political systems for their temporal democratic credentials. This would guarantee to dramatically shift our understanding of democratic politics and would begin to close the credibility gap between current (Liberal) democratic political practice and the popularly held democratic ideal. It would be a politics of temporalised democracy.

Temporalised democratic politics entails an explicit acknowledgement of the complex interrelation between plans, ideals and visions of the future as guides to political action and the foreclosure and reduction of future options as a result of actions and decisions taken in the past. Made explicit and acknowledged as dominant forces in decision-making, these supra-parliamentary influences would then need to be opened up to public scrutiny and influence. The livelihoods, safety and rights of non-voters past, present and future would have to be considered to the extent that these are implicated in and affected by the decisions and policies of the present. More importantly, responsibility and accountability have to extend to the future which is being created in the present and to the time–space distantiated 'others' whose lives are affected and whose options are foreclosed by the past actions of predecessors. Understanding of the complex interpenetrations and conflicts of the multiple rhythms of social and natural processes with industrial activity, economic reasoning, Newtonian science assumptions and Liberal Democratic ideology is essential to a meaningful assessment of the issues and the problems. Such a timescape perspective in conjunction with temporally extended responsibility would necessarily shift approaches to the environment from prediction and control to caution and precaution. Finally, intergenerational temporal equity, temporal rights, negotiations over temporal conflicts and their arbitration would be integral to a temporalised democracy.

So far, Liberal Democracies show little sign of such temporalisation beyond glimpses of it in the Rio Declaration and Agenda 21 of UNCED (Quarrie (ed.) 1992). Instead, it is citizens who have begun to institute changing temporal attitudes, approaches and visions. From the *Slow Food Association*, to consumer action over France's nuclear policy, animal liberation concerns and Greenpeace campaigns, citizens are re/formulating the temporal agenda: implicitly aware of the temporal conflicts created by the industrial way of life, they seek to retune their daily rhythms, take on board a long-term time perspective, take note of the spatio-temporal complexity of environmental processes and demand more appropriate, cautious and precautious action. Rifkin (1987) talks of 'time rebels' and identifies their approach to collective life with a temporal politics of empathy, participation and responsibility. Once we recognise our world as inextricably interconnected, understand nature as an extension of self and cultural activity, and appreciate the need to combat governance by money and science, such time politics become rational and the temporalisation of democracy desirable.

*Plate 4* Trees are us – protesters against the Newbury Bypass from *Resurgence* by Andrew Testa. *Source*: the photographer.

# 4

# INDUSTRIAL FOOD FOR THOUGHT

## For everything there is a season and a place

## Introduction

Food scares are the order of the day: salmonella in eggs and chickens, scrapie in lamb, BSE in beef and beef products, pesticide residues in fruit and vegetables, hormone-disrupting chemicals in baby milk – the list is far from complete, indicating merely the range of food hazards that have arisen in the UK since 1988. The increasing incidence of these health threats from food is indisputable; the scale of the problem impressive. Each one of these hazards is associated with the industrial production of food on farms and in the chemistry 'kitchens' of large corporations. Each one is invisible. All of them are time–space distantiated, thus difficult to tie down, establish causal connections, secure scientific certainty. Helplessness marks the reactions of the public, food producers, politicians, and scientists. The bewilderment, as I show in Chapter 5 with reference to BSE, is of course expressed very differently in each of these social spheres: politicians instigating one panic measure after the other; scientists disagreeing with each other and contradicting themselves in public; food producers worrying about their livelihoods, anxiously awaiting guidance and financial safety nets; consumers turning cynical or resigned, assigning blame to everyone but themselves and oscillating between abstinence from meat eating, opportunity buying and shutting the problem out altogether, wishing that it would go away, hoping that the timebomb was ticking in someone else's body.

With food more than with anything else we have relied on our senses to establish danger and safety. Throughout human history, we depended on the senses to tell us whether or not a food that is known to be edible was still fresh and fit for consumption. Today, we still rely on the nose to tell us whether food is fresh, ripe, or going off. From its look we can tell whether a piece of meat or vegetable is in prime condition, beyond its best or beginning its process of decomposition. A little squeeze can indicate the age of a food and the difference between staleness and freshness. Sight, touch, smell, even taste, however, are of no help in establishing whether or not any of the above hazards are present in an otherwise edible food: meat from animals infected with BSE or scrapie does not turn green or slimy; it does not look or smell any different from uncontaminated meat. Pesticides, fertilisers, herbicides and growth hormones do not show themselves in any way. The senses are effectively sidelined and for our safety we are dependent instead on the anonymous power of scientific knowledge and money. That is to say, only science, the knowledge system that has created the hazards in the first place, is able to identify their presence and, second, only the corporations who obtain their wealth from producing, marketing and distributing them have the necessary financial power to instigate world-wide monitoring systems and to develop the means to counterbalance the consequences, that is, chronic ill-health, the long-term prospect of pathology and disability and the fatal diseases that arise from them.

Even if the highly unlikely situation of corporate 'responsibility' were to materialise, that is, take precedence over the pursuit of money, the fact that currently 100,000 chemicals are on the market and 1000 new ones are being developed every year turns into a farce any research that seeks to link particular chemical products to specific diseases, fertility dysfunction and allergies. Both the scale of the problem and the traditional methods would mitigate against the establishment of meaningful results: abstraction from context, de-temporalisation, linear causality and proportionality, as I have argued in previous chapters, are the wrong tools to engage with these time–space distantiated threats, the inappropriate means to decode the *Wirkwelt* of these hazards. With respect to pesticides alone – that is, chemicals that are *designed* to be biologically active in the environment – the task would be insurmountable. As Colborn *et al.* (1996) report,

> The world market in pesticides amounted to 5 billion pounds in 1989 and included sixteen hundred chemicals. . . .
>
> Today the United States uses thirty times more synthetic pesticides than in 1945. In this same period, the killing power per pound of chemicals used by 900,000 farms and 69 million households has increased tenfold. Pesticide use in the United States alone amounts to 2.2 billion pounds a year, roughly 8.8 pounds per capita. . . .
>
> In 1991, the United States exported at least 4.1 million pounds of pesticides that had been banned, cancelled, or voluntarily suspended for use in the United States, including 96 tons of DDT. These exports included 40 million pounds of compounds known to be endocrine disrupters.
>
> (Colborn *et al.*, 1996:138)

Invisible, thus undetectable for ordinary citizens, and vigorously marketed by their producers, these and the other manufactured environmental hazards have become an integral and ineradicable part of our lives.

This raises the question of how to do battle with invisible enemies, how to combat what is beyond the reach of our senses and how to ensure that those who hold the keys to our well-being act to safeguard the livelihoods of current and future citizens rather than shareholders' profits. When blame is to be apportioned about food hazards, farmers are always the first port of call. While there is no denying that modern farming methods are as much life-threatening as they are life-giving, the story is a complicated one. It encompasses globalised economic, political and scientific activity and interests as well as taken-for-granted habits of mind, assumptions and practices. Moreover, time is centrally implicated in the development towards the contemporary state of affairs. Thus, the temporal issues involved provide clues as to why the move to non-chemical food production might be such a difficult change to achieve. Yet, time tends to be disattended in environmental analyses. A timescape perspective, therefore, sheds new light on these extensively discussed issues and allows us to pursue new questions and options for the future.

In this chapter, I outline some seminal changes that have taken place in farming practices and contrast these with the current role of industrial agriculture in the global food system. I then put a temporal inflection on industrial habits of mind in farming and on the implications of that approach for the environment. I follow this with an exploration of some of the timebombs ticking in the background of the industrial way of producing food. With an eye to the future, I consider options for change and scope for alternatives in farming practices, consumer choices and inter/national policy.

## The role of agriculture in the global food system

Every food system – defined as the totality of tangible and intangible means employed by a given human community for the production, conservation, distribution, and consumption of food – has profound effects on the environment.

(George, 1984:19)

Thus, the landscapes of Europe, for example, have been formed by human land use from the first wave of deforestation during the Neolithic period to the planting, extraction and replanting of forests during the industrial era. Even the treasured pine forests of Germany are the result of human activity, mono-culture replacements of deciduous mixed woods which used to cover much of that countryside in its ancient past. Across the world, large tracts of forest have given way to agricultural land for the production of crops and the grazing of animals. Irrigation without drainage resulted in salination of the soil and the production of deserts. Domesticated animals have replaced most of the wild herbivores that used to roam and graze the open land. Canals, ponds, reservoirs and dams have supplanted many ancient wetland areas. 'Our physical surroundings' as Susan George (1984:20) points out, 'can thus be "decoded" as incarnations of culture'. That is to say, the landscape is as much a record of agri/cultural activity as specific cultures are characterised by the degree to which they transcend the physical limits of time and space through knowledge and belief systems, communication, mobility, the storage, preservation and distribution of food, and the time–space distantiation of their actions (see also Chapter 1).

Before the eighteenth century, suggests Lynne White Jr. (1962), nine-tenths of most populations were involved at some level with agriculture and the production of food. This meant that changes in agricultural conditions and production affected the entire membership of societies. Today, less than half the world's population is working on the land, in Northern European countries it is a mere 1.5 to 3 per cent. The influence of the food system on society, however, is far stronger than these figures would suggest. Farmers may have lost their leading social roles to the agri-businesses that supply them with chemicals, seeds and feeds and to the processing companies that transform much of the produce of agriculture before it reaches the consumer, but the system as a whole is firmly integrated into the global economy from where it affects the lives of the majority of peoples on this earth. In this globalised context, local and largely autonomous agricultural production of food has given way to a de-contextualised, transnational system that prioritises export and trade over self-sufficiency and food quality with all the consequences that follow from this: live animals being transported vast distances within and between countries, fresh fruit and vegetables having to ripen in transit, meat and fish being preserved in ways most conducive to extensive transportation, mono-culture being developed and grown for the purpose of bulk movement. With respect to the environment this shift in emphasis means high energy use, widespread pollution, and dependence on chemicals with resulting environmental degradation and the kind of health hazards I write about in this book. Last, but no less pertinent, it has brought about the agricultural community's loss of the ownership of the means of reproduction, an issue I will return to below.

Analogous to the transformations of the landscape, the timescape too has been dramatically shaped by the way people provided for their sustenance and the way they related to the soil, the land, their animals and plants. While there are marked historical and cultural distinctions we must not conceive of these in either–or terms. There is no 'then and there' *vis-à-vis* a 'here and now' since the past and 'the other' are always present in the here and now, working from a distance and within: comfortably and/or conflictually, explicitly and/or implicitly, as reminders of the past and/or guides to the future. And yet, of course, there are some important distinctions to be made that irreducibly differentiate the industrial system of food production from others prevalent in distant times and places, distant that is from a Northern European, late twentieth century perspective.

## Agricultural timescapes: once upon a time and distant places

Once upon a time and distant places it was and is considered 'natural' for the seasons, climate, weather and location to impose limits on human activities, just as it was/is taken-for-granted that these restrictions pose a challenge to human ingenuity: how to store and preserve food through the unproductive part of the year and bridge the gaps between periods of scarcity and plenty; how to safeguard the reproductive capacity of land, plants and animals from one year to the next and for generations to come. Despite their quest to overcome the vagaries of the weather and to transcend the climatic extremes of the seasons, people were/are embedded in the light and dark, wet and dry, cold and warm, growth and decay, birth and death cycles of nature's earthly rhythms. Their future was/is 'laid out', not in detail but with an overall level of foreknowledge that allows people to conceive of themselves as integral parts of a continuum of past and future, predecessor and successor generations. This future which could/can be anticipated but not predicted, allows for planning ahead and for securing their own and successors' livelihoods. In those 'distant' times and places, time was/is not conceived independent of things and processes; rather, it was/is understood to be integral to them. For everything there was/is a season and a place. People, plants and animals were/ are known to be part of the great networked chain of being, creating for each other the conditions of life and death: from earth to earth, ashes to ashes, water to water, exchanging air.

This was and is a realm of co-evolution where, irrespective of the sub-systems' multiple and mutually implicating impacts, the system as a whole is able to absorb and recycle the outcomes of those activities. In other words, this is the eco-sphere where one sub-system's dissipated energy constitutes another's source of life, each one coordinated and reflexively synchronised with the entirety: contextual, contingent, transient and creative, each cycle a repetition not of the same but the similar, each therefore a creative rhythm of renewal. In slow co-evolution the earth system has been evolving whilst recycling its basic materials – earth, air and water – without producing poisonous 'waste'. In this way, and despite major disjunctures and discontinuities, James Lovelock (1979, 1988) and contemporary geophysicists assure us, earth as a living system has evolved and maintained its dynamic stability for millions of years within very narrow tolerances of temperature, atmospheric conditions, and concentration of salt in the oceans. (For an excellent overview of 'state of the art' knowledge in the geophysical sciences, see also Davies, 1996.)

Irrespective of the fact that human agriculture changed the face of the landscape, therefore, the earth system as a whole was able to maintain the necessary conditions for life. As long as the human production of food remained seasonal, which also means predominantly local or at least regional, and as long as the primary producers of food retained control over the means not just of production but reproduction, the system remained one of contextual, embedded, interdependent growth cycles. The crucial issue, therefore, is not whether or not farmers interfered with nature by domesticating animals, whether or not people needed to move on after an area's fertility had declined as a result of their activity, whether or not they tampered with growth cycles and maturation processes through their animal and plant breeding programmes, whether or not they controlled the processes of decay through methods of storage and preservation of seeds and food. It was rather that agriculture became abstracted from the ecological cycle of give-and-take and began to rely on external sources for the creation of fertility and by doing so produced polluting effects that had, in turn, to be externalised. That is to say, a crucial difference to cyclical, no-waste agriculture is the externalisation at both ends (i.e. input and output) of the industrial agriculture process which, on the one hand, generates poisonous waste that cannot not be re-absorbed into the system as a source of life and fertility and, on the other, produces a scale of degradation that transcends local conditions, affecting the evolved balances of the earth system as a whole at eco-sphere, hydro-sphere, geo-sphere and atmosphere level.

Next, I therefore want to explore briefly some of the implications for the environment and human safety arising from the shift, first, in locus of control from the agricultural unit to inter/national politics and global finance, and second, from self-sufficient cyclical production to reliance on synthetic chemicals. Later on in the chapter I indicate some of the tensions that arise in industrial agriculture from the imposition of abstract time on the inter-dependent and contingent rhythmicities of life and elaborate the link between the assumption that time = money and the industrial production of food, its intensification of methods, valorisation of speed, control of time, and resulting hazards for human and environmental health.

### *Agriculture squeezed: between suppliers and processors, industrial values and life processes*

The farmers' lead role as decision-maker has gone. Farmers respond to others in the system who set the pace for change and the rules under which the farmers operate. These rules are set by national politicians and by the bargaining power of the different actors in the food system.

(Tansey and Worsley, 1995:85)

This, as I indicated above, has serious consequences. Government and business establish export and trade rather than national and local self-sufficiency as the indisputable goal: only the former will dramatically increase the flow of money and thus show up positively on a country's official measure of wealth, its Gross National Product (GNP). Political concerns determine agricultural policies and with it priorities, subsidies and taxation. Economic considerations define the relation of food production to the global market (see also Chapters 2 and 3). Once farming is locked into the ideology of this globalised economic system, it is not the principles of life and co-evolution or concerns about fertility and the long-term reproductive health of the land, but rather political and economic considerations that dictate the nature, pace and intensity of agricultural practice. Explicitly and by default, therefore, government and business are the dominant forces shaping the development of agriculture and food production and with it their collective effects on the environment and human health. Much of the talk about the power of farmers is therefore misguided. French farmers can serve to illustrate this claim. By far the strongest lobby in Europe, they are unable to impose their will, impotent in their endeavour to prioritise French food for the French nation. If their campaign were successful, it would dramatically reduce the energy deficit associated with the export of food across national boundaries and the globe.

If farmers were the makers of their own destiny, we can be sure that the threat to their very existence would not have progressed at such a pace and on such a dramatic scale. Susan George (1984:27) provides some figures for the US where four and a half million farms have been eliminated since 1930 and where by 1985 'over 60 per cent of all farmers working in 1975 will have disappeared'. This trend, which is paralleled in Europe, is not explicable merely by the fact that mechanisation has dramatically reduced the need for manpower; it is not about farm workers losing their jobs (which they have done in vast numbers during this century) but rather about family farms going out of business. In 1995 in Bavaria, for example, 36 farms closed down every day of that year (headlines and lead articles in *Münchner Merkur*, 2/1996). Their owners have been forced to give up farming and with it a way of life that has often been part of their families' tradition for many generations. Across the industrialised North, farmers lost their livelihood because their capital-intensive agriculture was on a downward spiral of increasing debt in a context of decreasing fertility and productivity. Farmers are finding that they have intensified agriculture to a point where no matter how hard they work and irrespective of the quantities of chemicals they use, profits are declining, which means they are no longer in a position to raise the money needed to pay for their production costs. In the highly mechanised, chemically assisted, capital intensive agriculture associated with industrial production, these costs have increased dramatically and now make up more than 80 per cent of the gross income (George, 1984:26), thus tying farms ever more tightly to banks and money lenders. We need to appreciate, moreover, that even the biggest farms and agricultural enterprises are tiny players in this actor network of banks, national governments and transnational corporations which cover both agri-chemical businesses and food processing giants such as Nestlé and Unilever. Squeezed between these national and global players whose policies and interests constitute the farms' political and economic boundary conditions, what chances do they have for survival? The differences in power and room for manoeuvre can be illustrated with reference to the divergent ways with which the various actors in the food system approach their respective risks, that is, the way they deal with the potential for disaster and the uncertainties of the future.

### Taking risks: farmers, traders and limited companies

Farmers conventionally handle/d the risks of weather, seasonal variation, disease and pests by growing a wide variety of crops and keeping a range of animals so that failure of one could be offset by the stability and success of others. Grain and seeds were/are kept for several years ahead so that failing crops due to lacking or excessive rain, cold or sun could be bridged from one year to the next. Thus writes a French colonial inspector of the Upper Volta to his government in 1932 about the change in this rural community's livelihood:

> One can only wonder how it happened that populations ... who always had on hand three harvests in reserve and to whom it was socially unacceptable to eat grain that had spent less than three years in the granary have suddenly become improvident. They managed to get through the terrible drought-induced famine of 1914 without hardship. (Although their stocks were depleted, they were soon able to reconstitute them, at least until 1926, a good year for cotton but a bad one for millet.) Since then, these people, once accustomed to food abundance are now living from hand to mouth.
>
> (Quoted in George, 1984:22)

This is a story that is still repeated across the Third World wherever industrial agriculture has been adopted and where mono-culture cash crops have replaced indigenous crops and agricultural methods.

Under the industrial regime, risks are dealt with in a way that is unique to that system: a dramatically *reduced* range and variety of crops and animals are raised to enhance rationalisation, calculability and predictability. Fertilisers are to even out variations in the productivity of soil and crops. Pesticides and herbicides are to eliminate any unwanted animal and plant matter. Insurance is to provide financial compensation for a limited range of disasters. Instead of the traditional farmers' reliance on variety and diversity as safeguards against catastrophe and misfortune, the industrial way is to eliminate uncertainty by seeking to control both the physical conditions and the processes of industrial agriculture. This, suggest Tansey and Worsley (1995:90), seems inevitably to lead to an intensification of production, resulting in increased use not only of technological and chemical aids but also of energy and externally supplied, science-based feeds, semen and seeds. Industrial farmers therefore need skills different from those essential to the cyclical, no-waste form of agriculture. Instead of needing to acquire extensive knowledge based on experience and tradition, industrial farmers need to stay ahead of their times and their competitors: they need to keep abreast of the latest scientific developments in genetic engineering and chemical aids, embrace the up-to-date innovations in mechanised precision farming. They need to know about equipment maintenance and the use of a wide range of chemicals as well as attain accounting and marketing skills. Most importantly, they need high levels of bureaucratic skills in order to be able to deal with the tremendous amount of paperwork necessary to take advantage of the numerous (and continuously changing) subsidies and policies designed to even out the risks of modern farming. A second way of dealing with risk under the industrial scheme is through an economy of scale based on the assumption that bigger is better. The larger the outfit, it is argued, the more it can be rationalised with the aid of machines and the more likely it is for the bank to lend money in times of need. Moreover, as size is related to wealth, larger farms are better able to service their debts.

Since gluts and lean periods increase the insecurity and instability of agriculture, the industrial response is to seek to eliminate these variations in both the conditions of agriculture and the prices farmers receive for their products. In Europe after the Second World War, the latter economic strategy was pursued at the political level through the Common Agricultural Policy (CAP) which was to become the basis for the creation of the European Economic Community. Its aim was to keep prices steady and guarantee farmers a market for their products (see Hewett (ed.) 1995:75–9; Tansey and Worsley, 1995, Chapters 3, 6, and 9). In conjunction with the ideology of liberalism and a common market (see Chapter 3) this third strategy for averting risk had a number of highly problematic effects that increased rather than eased the economic pressure on farmers and agricultural food production: it intensified farming, produced tremendous surpluses, progressively reduced farmers' profit margins while extensively increasing the profits of agri-businesses and food-processing firms, and it forced record numbers of farms out of business. It has not delivered the objectives for agriculture but has enriched traders, transnational corporations and banks. In other words, while it destroyed the livelihoods of a substantial number of farmers and their families it has worked well to enrich the other leading players in the food system, whose primary concern is not agricultural re/production but the activity of trade which creates both the flow of money and the potential for profit. Judging by the results, this triple risk strategy does not bode well for either agriculture or the environment.

One group of actors that are doing well out of the industrial farming system are the food traders – the importers, exporters, brokers and merchants – who trade agricultural products on the global market. Two things are important to appreciate here: first, that the food trade is dominated by a few transnational corporations. With respect to grain, for example, six companies dominate the world's trade in that commodity. For the entire food trade it is a mere 15 companies that account for about 80 per cent of all trade. Second, *for trade in food, unlike for agriculture, the cost of a product is not crucial to wealth creation. Instead it is trade itself that is of importance: the more a commodity changes hands the better.* Since the demand for commodities rises and falls with a number of predictable and unpredictable factors, the traders' prime concern, like that of the farmers, is to counteract that variability and future uncertainty. Their solution to handling risk, however, is very different from the strategies adopted by traditional and industrial agriculture. In response to their risks, traders have developed the 'futures market': they trade in the future prices of products for which there is/may not yet be a market or at least not yet an optimal market. For them, the future equals money.

The futures market is a commitment to buy/sell something of a specified standard, at a pre-appointed time and at an agreed price. To engage in this 'hedging', traders do not have to be in possession of that commodity. Rather, they trade futures for products in the hope and with the expectation that this contract can be sold on for profit before or shortly after the time is up to fulfil the commitment. This means that the same commodities may be bought and sold many times over. Such futures trading happens within the regulated context of some 75 organised and electronically linked exchanges which are located in the major financial centres of the world. Moreover, this is not a new way of dealing with the uncertainty of trade in the food industry but, as Deirdre Boden (1997) explains in a fascinating paper on futures trading, it started with rice in seventeenth century Japan. Writing about the high-speed speculations of contemporary futures traders, she suggests,

> These young traders are trading in time itself, which is to say, in the momentary forward fluctuation of price and value. The latter are, by extension, expression of the most abstract sort: of money itself and, even more abstractly, of the price of money at some future point in time, in other words, in future interest rates.
>
> (Boden, 1997:14)

> The future value of a commodity contract is not simply a matter of linear time – increasing or decreasing in some patterned way – but is linked instead to complex multivariate calculations: inflated, discounted, hedged and even expressly devalued.
>
> (Boden, 1997:15)

The traders' power resides in the scale of their resources and their access to finance, in the integrated and networked nature of their operation and, finally, in their ability to manipulate not just the market but the future. Importantly for the issues I am discussing here, as Tansey and Worsley (1995: 108) note, futures trading 'is done by technical traders who do not and never will actually deal in physical commodities but will only speculate in the market'. The livelihoods of farmers, soil fertility, and/or environmental sustainability, therefore, clearly do not feature within the framework of their professional concerns. This means, there exists no point of intersection, let alone overlap, where it could be argued that the interests of agriculture, trade and the environment converge, except that is, for their collective need to earn money through their activities. Thus, the future as money, as speculated upon economic potential, is difficult to reconcile with a future that is to sustain succeeding generations of humans and other species and a future that is to harness the survival of the earth system as a whole.

Food processing companies, a third group of actors in the food system, have yet another set of concerns and thus different ways again of dealing with risks and the vagaries of the future. Since I have dealt with some of these in the previous chapter, I shall only raise a few additional points here. First and foremost, with the creation of limited liability companies, risks can be externalised, which means they are not borne by individuals and corporate entities but spread across society who has to pick up the bill for and shoulder the socio-environmental consequences of corporations' actions. A second way to minimise corporate risks is to externalise them to agriculture, to pass them on to the farmers who supply food processing companies with the raw materials. By stipulating tight contractual conditions about the time of delivery, the quantity, size, colour and uniformity of the product, and even about growing methods and chemical regimes to be used, processing companies have plenty of means to pull out of a contract, should the need arise, since failure on one of these conditions would constitute a breach of contract. Third, and as I have already suggested in Chapter 2, transnational corporations can take their custom anywhere in the world, that is, wherever they can secure the best deal for themselves: for a global operation there is always summer somewhere and always one agricultural producer willing/able to outbid the rest. Finally, to guard against the potential of having the polluter-pays-principle activated against them, companies play a game similar to the card game where a designated 'bad card' is secretly passed on at great speed so that the players avoid being caught with it at the end of the game. This means that, in order to escape prosecution, companies engaged in risk-intensive socio-environmental practices get sold and resold many times and at great speed, as was the case with one of the main rendering businesses involved in the BSE crisis (see next chapter).

From this very brief look at the different ways in which some of the main protagonists of the industrialised food system relate to the uncertain future, we can see that farmers clearly occupy the riskiest position. Not only do they have the least effective means of dealing with the hazards they face in the course of their work but, more significantly, they are squeezed between capricious policies and the interests of traders and food processing companies who, in turn, even manage to offload some of their risks back onto agriculture. Alongside the other contributors to the food system, farmers are stuck in an ideological system, a habit of praxis that undoubtedly serves big business but whose credentials are far more dubious with respect to the well-being of agricultural and rural communities across the world, the consumers of food, and the environment.

In public discourse, the industrial way of intensive farming is considered to be the 'conventional' agriculture. With its dependence on mechanisation, chemical agri-business and food processing companies, however, the industrial method of farming has been fully institutionalised only since the middle of this century. There is, therefore, nothing conventional about it. Rather, it is historically a very recent and rather troubled development in agricultural methods: high-tech, capital intensive, labour saving, with intensive crop production and animal husbandry, and a linear system of chemical input and poison-waste output. It is, moreover, a system that does not inspire confidence about its long-term future prospects. Industrial farmers as a social group, so we are frequently told in the UK news, have the highest suicide rate and their health records leave a lot to be desired. The 'State of the World, 1994' Report of the Worldwatch Institute reveals a higher than average rate of 'certain cancers among farmers and agricultural workers who are exposed to a wide array of natural toxins and synthetic hazards' (Misch, 1994:123–4) and gives a long list of cancers associated with industrial farming and the range of increased risk. This report argues a link between chemical pollution and suppressed functioning of the immune system and it suggests that 'farmers might be the vanguard of a population at risk of cancers linked to environmental pollutants . . . farmers may be the "canaries" of environmental health' (Misch, 1994:125). Industrial farming comes with a big price tag. Yet, it is clearly not enough to state this, not even sufficient for farmers to recognise this. The steep rise over the last decades of ecological criticism of the system has not (yet?) had a major impact. Some of the reasons for the failure to change direction I have already indicated in this and earlier chapters; others are still to follow. Industrial farming in its current form may be a fairly recent phenomenon but its roots reach to the depth of (Western) post Enlightenment, industrial, scientific, liberal democratic and economic being. Here, as in the other areas of social life I discuss in this book, its taken-for-granted, invisible temporalities are deeply implicated in that system and it is therefore important that they are rendered explicit as part of the process of battling with invisible enemies.

## The timescape of agriculture under conditions of industrial production

### Industrial habits of mind revisited: when nature and time are money

Food production, the industrial way, is guided by a set of assumptions that we have already encountered in previous chapters: nature and time are conceived in terms of money; both are viewed as resources to be exploited and/or commodities to be controlled and sold. In Chapter 2, I indicated in a general way what might flow from this perspective and approach. Here, I briefly want to revisit this argument with special attention to agriculture. I refocus existing debates by placing a temporal emphasis on the contrasts and similarities between industrial and organic ways of producing food, on their approaches to 'productivity' and 'efficiency', decontextualisation and standardisation, on control and the pricing of products. First, we need to remind ourselves that the economic view of nature and time comes as a package of interconnected and interdependent taken-for-granted assumptions that are rooted in the linear perspective, Newtonian science and neo/classical economics (for detailed elaboration, see Chapters 1 and 2). In addition to the economic bias, key features of the package are a tendency to standardise, simplify and decontextualise; a disposition to think in either–or terms and to create 'others'; an almost religious zeal to quantify and achieve control; and a blindness towards below-the-surface processes and all things temporal. In agriculture, this clustered habit of mind finds once more a unique expression. A substantial part of my argument and some of the data presented in this part of the chapter are based on work arising from the *Time Ecology* project at Tutzing, Germany – most specifically the eighteen months project on agriculture which culminated in a conference in May 1995 and a publication in the autumn of that year (Schneider *et al.* 1995). (For information on the project more generally see Adam *et al.*, 1997b and the introduction to this book.) With respect to this project's findings, I draw most extensively on papers published in Schneider *et al.* (1995).

The tendency to decontextualise actions, processes and phenomena takes the form of seeking to overcome seasonal variation as well as differences in locale and between individuals. This rationalisation of rhythmic and variable processes is to facilitate quantification and measurement, predictability and control. As with any other industry, the standardised time of clocks and calendars determines routines and serves as abstract exchange value. In dairy farming, for example, feeding, milking, cleaning, even the timing of reproductive activities are largely regulated with reference to clocks and calendars. Thus, the temporal pattern and timing of milking – the how often and when – bears little resemblance to its original function. Where calves left with their mothers would suckle about ten times a day, milking (mostly) happens twice a day. Equally, calves tend to be fed on a twice daily schedule which can adequately cover their nutrient requirements but leaves them with a shortfall in suckling and comfort, a deficit that tends to materialise in the form of behavioural abnormalities, stress and reduced resistance to illness.

The ideal of industrial production is for everything to be standardised. This means daily rhythms designed to secure maximum efficiency of operation. In agriculture, more than any other industry, however, there are limits to the extent to which seasonal and daily variation can be rationalised to conform to a decontextualised and de-temporalised standard. In other words, abstraction from context and standardisation, both successfully achieved in other industries, are not so easily imposed when the produce are living entities, when the principles are applied to the growth of plants and the maturation of animals. Plants and animals are ineradicably tied to the rhythmicity of nature and the cosmos: their physiology is determined by it. Their maturation processes are tied to it. Their reproductive cycles oscillate in synchrony with it. Their growth and decay patterns are guided by it. Intense bursts of growth during spring and summer are followed by times of decay or rest, inactivity and recuperation during autumn and winter.

Despite this inescapable tie of (almost all) life forms to the earth's rhythmicity and seasonality, attempts are being made to transcend that context-bound temporal characteristic and incorporate agriculture more fully within the industrial way of doing things: improve productivity, efficiency and profitability by creating sheep that 'produce' lambs twice a year, crops that grow in extended seasons, apples that keep for an extra couple of months. In each of these cases nature is considered to have far too many and long stretches of non-exploitation and idleness, 'unproductive' periods during which it could be earning money and making a profit. 'Genetic boost makes our Cox's crisper' boasts a headline in the *Independent* newspaper (Byrne, 1/9/1996:8). The article reports on research that changes the DNA of Britain's favourite apple in order to be more competitive in the global market of apples. The British apple's shortcoming is that it only keeps (in sellable condition) from autumn to March, after which time only foreign apples are to be found on shelves of UK shops. The obvious answer to this economic problem was therefore to redesign the British apple to 'stay fresh' until the next crop of British apples are ready to be sold in our shops. The idea that one might eat some other seasonal fruit or vegetable is not under consideration: that would be a retrograde step that adds neither to the country's GNP nor to the scientists' international reputation. An apple that keeps for a whole year, in contrast, not only keeps foreign imports down but its superior design is also expected to boost British exports. To the joy of the scientists responsible for this research, 'tests have already shown that the genetic changes introduced into individual trees are reproduced in future generations'. Concerns about 'meddling with nature' are brushed aside with the argument that after all, Mr Cox created the British cox when 'he crossed two varieties of apple to produce this wonderful fruit in the first place'. I am considering the difference between traditional plant breeding and genetic engineering in a later chapter. Here I merely want to illustrate with this example, the quest for eliminating seasonal variation and behaviour. The underlying expectation is the availability of all foods, anywhere, anytime: permanent harvest season in every supermarket across the globe.

Industrial social relations, as I showed in part one of this book, are inextricably tied to an approach to time as money. This time = money permeates not just all employment relations and, on the basis of this, the price of a product, but it applies equally to the complex temporalities of the product. In agriculture, it is relevant to the way food production processes not only intersect with these temporalities but shape and control them with reference to the need of production. Thus, it matters how long it takes for a crop to grow and ripen or how fast an animal reaches its optimal weight for slaughter. The economically focused breeding of pigs, for example, has dramatically reduced the time it takes for the animals to reach their ideal weight for slaughter. Helmut Bartussek (1995:67) provides statistics showing that in 1800 it took two to five years for a pig to reach a slaughter weight of 60 kg. By the beginning of this century, it only took 11 months for the pig to reach a weight of 100 kg. Today that same weight is reached before the pig is half a year old: ready for slaughter before it has lost its baby teeth – clearly a massive improvement in productivity! It matters similarly to what extent the temporal processes of ex/live food stuffs can be controlled in transit and storage by means of preservation and disembedded ripening: salting, drying, canning, freezing, irradiation and genetic engineering are some of the ways to achieve such control over the temporal processes of life, over growth, ripening and decay. Finally, it makes a crucial difference whether efficiency and productivity are associated with maximum yield in the shortest possible time or with a measure of output over a lifetime. Let me expand on this last point with an example of milk production and follow it with some expressions of my puzzlement about the mystery of food prices.

Under an industrial scheme of agriculture, the productivity and efficiency are understood in a very specific way, that is, with reference to individualised and particularised performance which is measured in abstraction from the long-term perspective of potential lifetime performance. This means, the 'productivity' of a milking cow is measured not over her lifetime but, rather, by how much milk she gives per day, month and annum. Figures provided by Günter Postler (1995:59), show that an average Northern European industrial milking cow will give some 5–6000 kg of milk per annum for a duration of about three years (three lactations) which compares with 800–1400 kg of milk per calf of non-domesticated animals. Under the industrial system of milk production, however, she tends not to reach even one quarter of her potential age. Instead, she is worn out at the age of 5 after her intensive period of continuous pregnancy and lactation. In the US her productive period is down to 2.2 years and in Israel it is reduced to 1.8 years. The outbreak of BSE in the UK is taking this downward trend even further. In an effort to combat the infection, the replacement period is down to 18 months. The BSE crisis apart, productivity here means increasing output over ever shorter time spans: the faster something goes through the system, the better it is for the economy and a country's GNP. This is the economic valorisation of intensity of use value in the immediate present, measured in decontextualised units of performance, imposed on an exploitable resource that happens to be a living creature.

Cows reared for extensive, lifetime productivity have a distinctly different time profile. Compared with their short-lived, intensively reared sisters, they bear substantially more calves – the record being 39 calves by an Irish cow who lived to the age of 48 – and provide the farmer with about the same amount of milk over the period of their natural life-span. At the same time, they need less additional feed and medicines. They will suckle their own calves, passing on not just contentment but disease resistance. On the basis of these combined factors, the risk that any of these extensively reared and organically fed animals or their offspring might carry the BSE agent is exceedingly low. The quality of both milk and meat are unquestioningly superior and less likely to pose a health hazard to humans. Thus, depending on the time-frame of the perspective and whether or not the future is discounted (see Chapter 2 for this economic practice), productivity and cost-effectiveness are defined in opposite terms. From the short-term, economic perspective of intensive agriculture, productivity and cost effectiveness are tied to speed of production and to control over the ageing process. From a longer time perspective, the extensive way of lifetime production combined with a minimum of costly external input is the most cost effective.

The price differences between the products of industrial and organic or traditional farming are conventionally explained with reference to governmental subsidies, economies of scale, low labour costs and the fact that the true cost of energy is not factored into the product price of industrial farming. Energy costs are externalised, which means that their full costs and consequences are carried by the glocal community in order that consumers, wherever and whenever they are, may enjoy the semblance of permanent harvest festival time. The four factors undeniably play a crucial role in explaining the mysterious price difference between locally produced and world-travelled, jet-setting food. They provide some answers to questions about how it is possible that a cheese, blended from milk that comes from sources across Europe, can be cheaper than cheese made with milk from local cows and goats; how it can be that frozen lamb imported from New Zealand is cheaper than the locally bred lambs that populate the Welsh hillsides by the million; how it could be that apples from South Africa are half the price of apples grown by the organic farmer on the edge of the town. Undoubtedly, it is cheaper to produce apples in vast quantities and employ the cheapest possible labour to harvest and pack them. But then, are not the savings in labour costs and economy of scale offset by the cost of fertilisers, pesticides and herbicides and by some of the high costs of mechanised production? When the savings from cheaper labour and large-scale production are offset by the high costs of machinery and chemicals, it becomes difficult to see how, on top of this, the export costs are being absorbed so as to result in prices lower than local produce. Economists have not provided convincing explanations of how produce such as South African apples can cost less in the UK than locally grown apples, given that they first have to be transported to a harbour, get handled and loaded onto a ship, transported to a port in Europe, unloaded and transported to a central market, and get traded at a variety of levels from the global futures market to the local shop. Even if we allow for the externalisation of energy costs to society, there is still much that should add to the price of these apples. What makes the price difference difficult to fathom, in other words, is not just the enormous amount of energy involved but also the fact that at each single step along the way somebody is making money, which means the longer the chain, the more people are involved earning a living and making a profit from the product. For me, therefore, the four factors – subsidies, low labour cost, externalisation of energy costs, and economy of scale – do not shed sufficient light on the matter to solve the mystery.

I remain puzzled. What is it that makes the price difference between industrial and local and/or organic products? Are we confronted here by the hidden price for health and quality? Are we encountering the food equivalent of the handmade Swiss watch whose high price expresses an appreciation of quality in terms of physical labour and skill in combination with care which takes time? Are we in the throes of a post-food-scare re-valuation of time = money that works according to a rationale different from the classical economic one? We certainly seem to be experiencing a reversal or at least a dismantling of well-established values such as the ones noted by Christine von Weizsäcker when she suggests that food products gain in status proportional to the distance they have travelled: only foodstuffs that have travelled for more than 1000 kilometres could possibly count as delicatessen (quoted in Schneider, 1995:8). It seems to me that this view no longer holds. Today's delicatessen is not the undated honey blended from nameless sources across the globe, nor the frozen lamb imported half way across the world from New Zealand, nor the mass-produced bacon from Denmark. Instead, it is this season's Presili honey, the fresh Pembrokeshire lamb and the home-cured bacon from the local farm that command top prices. They encapsulate quality, wholesomeness and with it the hope for safety from hazards and threats to health.

### Operating in the geological timescapes of soil and water

Industrial agriculture, as I suggest above, is present-oriented and places emphasis on intensity of production. Its time frames of concern, therefore, are extremely restricted and made worse still by the economic discounting of the future. The time scales of effects of this form of agriculture, however, are vastly more extensive than those of any other agricultural system, its chemical legacy accumulating in soil and water for unknowable periods into the future. The unbridgeable gap between time scale of concern and effect is worsened by the fact that most of the deeper level regeneration of soil and water is possible only in geological time scales, that is, it too is outside the timescape of industrial production. Again, the Tutzing *Time Ecology* project is extensively involved in these issues and I therefore draw generally on this work and on two of the published papers (Kluge and Schramm, 1995 for water; Scheinost, 1995 for soil) in particular.

Soil is the time–space where biosphere, hydrosphere and atmosphere interact in a continuous interchange of nutrients, water and oxygen, as well as vegetation, organisms and micro-organisms. Soils are generated and they age in an ongoing dynamic of de- and regeneration. At some level, therefore, soil is a renewable resource. The issue gets more complicated, however, when we take into consideration that soil is regenerated at vastly different time scales. While some top soil can regenerate within 100 years, the clay subsoil takes some 10,000 years and the layer below that – even at the mere depth of one metre – takes 100,000 years to reach its optimum nutrient quality. So we can say that it takes between 10,000 and 100,000 years to re/generate one cubic metre of soil. This time scale of re/generation clearly places soil outside the human time scale of action and thus beyond the range of what meaningfully could be called a renewable resource. With a gap of that order of time between soil loss and renewal, this primary source of livelihood and survival must be of crucial concern now and for the future.

The first thing to note is that there is no viable long-term alternative to the soil and its nutrients as a basis for plant life as well as animal and human survival. The soil provides not just food and raw materials, it is also a converter of energy and minerals, and it is a store for nutrients, water and, by the same token, for pollution. The health and stability of the soil can be damaged by movement, such as water or wind erosion, by structural changes, such as compression or waterlogging, and by changes in the soil chemistry, such as those brought about by non-biodegradable pollution or salination from irrigation. Through industrial activity, logging and agriculture, human activity contributes to soil degradation at all these levels. Much of the damage, however, is occurring out of sight, in the *Wirkwelt* below the surface. Moreover, we are unlikely to encounter the bulk of the symptoms of degradations arising from contemporary industrial activity in our lifetime, since action and impact are time–space distantiated and marked by multiple time-lags. Instead, industrial societies externalise the problem temporally to future generations.

Even less visible and accessible is the effect of the industrial way of life on the elementary resource water. Of course, we can see when the river in the valley below is turning yellow or purple-blue as a result of up-stream pollution, but very little of the long-term damage to the 'elixir of life' is visible in this way. We think of water as a renewable resource, one that replenishes itself from season to season. Sometimes, when winter rainfall is below average, we may find that the reservoirs have not been fully replenished, which means drought conditions are likely during a dry and hot summer. On balance, however, Northern Europeans, and people of the Northern Hemisphere more generally, expect water to be available all the year round: the water companies are not doing their jobs properly if water is running short during the summer. We are so habituated in the transcendence of seasonal variation – through work practices, global food supply, electricity to light the hours of darkness, and fossil fuel energy to heat houses in the winter – that we make no exception for water. On the contrary, during the periods of lowest 'stocks' we expect as a consumer right to have enough water for extra high usage. This habit and the associated expectancies in conjunction with industrial pollution have grave consequences.

Governments and water companies in their wisdom have decided that the answer to any water shortage is to bore deeper: when the resources in reservoirs, dams, rivers and canals run short, water is to be brought up from the depth of the earth. As a second measure they have imposed annual limits based on averages. The latter tactic, as Kluge and Schramm (1995:39–41) point out, is useless since averages do not take account of seasonal variation and ignore long-term changes such as global warming. They serve no other function than to secure the *status quo* and to ensure that water companies do not need to engage in long-term precautionary measures. The boring deeper strategy is equally flawed, if substantially more worrying. This is because, unlike the surface water which is a renewable resource, the ground water and the very ancient sources of water below are so slow to regenerate that from the perspective of human action they are non-renewable. As with the soil, we are talking of thousands of years of regeneration time and even then we cannot be sure that the water would regenerate to anything like its current pure and pristine state. There is no way of knowing whether or not these ancient fossil waters will be irredeemably contaminated with the chemical and heavy metal pollution of the industrial way of life. In other words, as we draw from these ancient sources of clean water, we deplete what is in human time-scale terms a non-renewable resource and we replace this not just with younger but with potentially polluted water. In Germany's main industrial area, the Ruhrgebiet, chemical and metal pollution has reached groundwater to a depth of 150 metres which is well below the level of most mineral water extraction. This pollution is likely to be non-retractable. As such, the problems of resource depletion and irredeemable pollution are temporally externalised to future generations of beings.

A brief recap of some of the temporal issues of relevance to both soil and water seems helpful here before moving on to the minefield of timebombs waiting to go off. Of central importance are contextual time–space and temporal variations: seasonal, from year to year, and long term. At human rather than geological time scales, much of the non-biodegradable pollution is non-retractable. Moreover, most of the damage occurs in the invisible *Wirkwelt* below the surface at time scales that are outside the range of economic and (most) political concern. The time gap between overuse and re/generation is unbridgeable. The problem is externalised into the future. Time-distantiation and time-lags between action and symptom constitute conditions of irreducible indeterminacy and therefore unpredictability. Both of these elementary sources of life are dynamic: soil and water change. They age. They re/generate and they are degradable. They are simultaneously resilient and vulnerable. The metaphor of the timebomb comes to mind when considering the hazard potential of the industrial mode of exploitation of these two primary sources of existence.

146

## *Timebombs awaiting their 'harvest'*

A healthy crop of hazards is good for the economy. Hazards show up positively on a country's GNP, on equal footing with export and war. They create jobs, keep hospitals, scientists and politicians busy. They facilitate the circulation of money. Financial traders are thriving on them: nothing like some good disasters to create their ideal conditions of volatility. The ceaseless inventiveness and innovative capacity of business means that hazards are appreciated as sources of new opportunities. From an economic and from a scientific perspective, we need to recognise, a healthy dose of hazards is an unmitigated good thing. We should not expect too much effort from these quarters, therefore, in the direction of preventing the damaging effects of industrial economic and scientific activities since they constitute one central source of their future wealth. Thankfully, humans tend not to be exclusively motivated by the logic of their institutions which means they cannot be relied upon to always act 'rationally'. There is much encouraging evidence from business and science that concerns for the long-term future, for safety and environmental health are motivating factors for action that spin the downward spiral in new and unpredictable directions. Such 'disloyalties' to the logic of industrial development and globalised economics needs to be fostered with all the means at our disposal. I make some practical suggestions to that effect in the last part of this chapter which follows a more detailed exploration of the timescape of industrial food.

At the end of this section I thought it might be helpful to start, without much comment, a list of impending hazards associated with the industrial production of food. I list them with reference to their potential time order of magnitude. To see them collated in one place helps to concentrate the mind. The gravity of the situation becomes visible when we see next to each other things we tend to shut out of our minds, dangers we hoped would disappear with stubborn disattention. What follows is merely a selection of largely invisible, time-distantiated hazards associated with industrial food production. Some of these may not carry for us in the present the status of hazard but, without question, they constitute hazards for future generations as well as long-term human survival. Readers may wish to extend this list with their own favourites.

### *Water*

- Long-term degradation, pollution, loss of clean drinking water, and reduction of available fossil water;
- Contamination of ancient waters with chemical and other industrial pollutants reaching a depth of 150 metres already;
- Accumulation of non-degradable pollutants;

- Damage to micro/organisms;
- A gap of thousands of years between unsustainable use and re/generation (ignorance in this field means the figures could be immeasurably larger or, indeed, smaller);
- A generational inequity extending to many thousands of generations.

## Soil

- Salination, desertification, erosion, structural damage, degradation and threatened loss of nutrient base for food;
- Long-term pollution and toxification;
- Accumulation of non-degradable pollutants;
- Damage to micro/organisms;
- A gap between unsustainable use and regeneration of hundreds of thousands of years;
- An intergenerational inequity extending to many hundreds of thousands of generations.

## Food

- Loss of nutritional value in processed foods;
- Vulnerability to disease in plants and animals due to reduction in genetic and species diversity;
- Unlabelled additions and structural changes such as irradiation and genetic engineering to processed foods;
- Invisible contamination of a radioactive, bacterial, viral, and genetic kind in soil-, water-, and air-based foods;
- Equally invisible contamination with pesticides, herbicides, and fertilisers;
- Hormone-disrupting chemicals in the food chain causing death, long-term illness and disability as well as damage to the reproductive, lymphatic, and endocrine systems in all species of mammals, including humans;
- Time-gaps between action and symptom ranging from instantaneity to a completely open-ended future: 0 for instant death by poisoning; 24 hours for salmonella; somewhere between 5–20 years for the BSE agent and some hormone-disrupting chemicals; up to thousands of years for radioactive contamination as it is passed on from generation to generation; a completely open future not just for the long-term effects of plastics but, equally if not more worrying, for the impact of genetic engineering experiments and activities.

## We are what we eat

In the long term we can only live when we allow things their own time, when we do not force them into a temporal scheme based on nothing but utility.

(Altner, 1993:136)

Our health and well-being are intricately webbed with that of the plants and animals we eat. Their stress becomes part of us. Their artificiality does not leave them before they are ingested by us. Their loss of vitamins and taste due to travel and storage is our loss too. It matters, therefore, what happens to them before they get eaten. This subject is widely and compassionately written about, yet the time dimension tends to be left implicit. Let me therefore add selectively to this popular and academic literature by pointing to just a few of the temporal issues of concern.

### *'For everything there is a season' (Ecclesiastes 3,1) and a place*

Not any more. Since export, movement/transport, and the creation of multiple trade chains are 'wealth-creating', thus crucially important from an economic perspective, our foods criss-cross the globe, busily adding to every country's GNP along the way. In order to suit this particular economic need, however, food had to be adapted and redesigned to suit the new conditions of its 'labour'. Where it used to be grown and consumed locally, the bulk of food today has travelled extensively. In transit, it had to ripen or be preserved in some form in order to arrive 'fresh' on the supermarket shelves. The consumer expects apples all the year round and strawberries at Christmas – so we are told. This and this alone, we are assured, is the reason why, as Joanna Blythman writes,

> The fruit and vegetable buyers for our large importers and supermarket chains spend their days chasing the sun around the world, in search of climates that can produce fruit and vegetables at times when they would be otherwise unobtainable or taste awful. Raspberries at Christmas, apples in June, and sugar snap peas in March.
>
> (Blythman, 1996:24)

Because she thinks that, regrettably, global sourcing is here to stay, she provides her readers with rudimentary geographical knowledge and tips on how to be sure that this world-travelled food is 'seasonal' in its source country and has not already spent several months irradiated in some store.

Freedom from context is one of the great achievements of science and industrialisation and for those with the necessary money it means emancipation from the cycles of want and plenty.

> Some 100 to 150 years ago, around 95 per cent of what people ate had been produced within sight of their church steeple. This had both advantages and disadvantages. One could see and experience how the crops and animals that one lived from grew and flourished. One was, however, much more dependent than today on *whether or not* they grew and flourished.
>
> (Schneider, 1997:86)

This transcendence of seasons and locale, this success story of industrialisation, as Manuel Schneider (1997) calls it, comes with a price tag: decreased diversity of kind and genetic material, decrease in vitamin and nutrient values, decrease in taste, coupled with an increase in hazards from chemicals, irradiation and genetic improvements. On the positive side, it has to be said that the products of globally sourced food *look* better: bigger and shinier, perfect in form, size and condition. Beauty, the pleasure to the eye, is to offset lack of taste and some of the other minor consequences for jet-setting visual perfection. To achieve visual shelf perfection, fruits such as apples are harvested green which brings a number of advantages to growers, traders and retailers. Because unripe apples are still hard, they can be taken off the trees mechanically, all in one go. Not yet ripe, they do not bruise easily when handled, and thus cope well with haulage and packaging, that is, without any of the normal tell-tale signs of careless treatment. This makes these world-travelled beauties not only a visual delight but cheap and, to top it all, they have an excellent shelf-life. 'What more could we possibly want?' ask the commercial growers, the traders, the wholesalers and the retailers.

Since not all varieties of apple, for example, are amenable to green harvesting, only those that are suitable to this treatment find their way into our shops. Observation will tell us that 'Golden Delicious' and 'Granny Smith' are the two varieties best suited to that scheme. Consequently they are available in great quantities, at low cost, all the year round. Six thousand named varieties of native apples, for example, are listed in the UK Apple Register; just nine of these constitute the bulk of UK apples that are grown commercially. Diversity of tastes, consequently, becomes the private preserve of people with gardens who inherited old trees or have had access to specialist nurseries who still sell old varieties.

What then is the exchange involved in the decontextualisation of food? By being freed from seasons and locality, our diet has become less monotonous at any one period of the year. Gone is the time when winter kale and sprouts were the staple 'greens' for what seemed weeks and weeks and weeks. We can have tomatoes, peppers, cabbage, lettuce and cucumber all the year round: no need to eat winter greens followed by spring vegetables and an abundance of summer vegetables and fruit. Today, we have the opportunity to eat the same variety of apple, tomato, pepper, and lettuce all the year round. The relative monotony of a seasonal diet has been replaced with the absolute monotony of all-year sameness and uniformity as well as a staggering lack of taste. Lost is not just the vast variety of kinds and diversity of their tastes but the joy that comes with the expectancy and arrival of new seasonal foods: the first peas, carrots, cucumbers, strawberries, apples. With the achievement of this all-year sameness and uniformity the meaning of freshness had to undergo some considerable changes.

## Counterfeit freshness

The Concise Oxford Dictionary has a long list of attributes that denote the meaning of freshness. Recent, not preserved, pure and untainted, not stale, musty or vapid, newly made are some of the characteristics most suited to the description of fresh food. In everyday language, 'fresh food' means food that has not been preserved by drying, salting, pickling, smoking, tinning, bottling or freezing. This understanding which contrasts freshness with preservation, however, leaves a large grey area, that of stored food. This food is 'fresh' in so far as it is not preserved by any of the traditional means but it is not fresh in the spirit of the word, that is, freshly harvested from fields and orchards. In addition to this difficulty, contemporary industrial methods of storage take the meaning of freshness into an altogether different realm.

The storage of food in cool, dark and airy places as a means of keeping decay at bay is an ancient method of bridging the gap between one harvest and the next. It is one of the many ingenious ways humans have invented to sustain their food supply during the dormant period between the seasons, between harvest and new growth. Each fruit and vegetable species and their individual varieties have their own time-frame within which the onset of decay can be delayed. In the course of human history, much knowledge had been accumulated about how to extend the period of available 'fresh' foods by cultivating different varieties so as to make use of their markedly differing ripening times, keeping spans, and patterns of decay. Moreover, since harvesting times and decay periods were part of common knowledge, consumers could work out with reasonable accuracy how old their purchased 'fresh' foods were likely to be.

Industrial food storage, in contrast, has moved away from an emphasis on variety towards an economy of scale where large quantities of the same foods are kept 'fresh' by means of a controlled-atmosphere ripening, chemical preservation, irradiation and genetic engineering. These processes and their timescapes of harvesting and ripening, storage and decay no longer form an integral part of public knowledge. Consumers therefore are no longer in a position to work out the age of the 'fresh' food they are buying. Without laws stipulating the public identification of the date of harvesting, on the one hand, and method of preservation used, on the other, consumers are left in the dark about the temporal history of the foods they purchase. This, surely, is a significant difference worthy of further public attention.

A brief look at some additional differences between traditional and industrial methods of extending the period of food 'freshness' will help to illuminate some of these generally disattended issues. In *The Food we Eat*, Joanna Blythman (1996) informs her readers of the contemporary range of methods used for keeping food in a state of simulated freshness, thus foregrounding for public attention a social issue that otherwise tends to be largely invisible.

> Fruit brought by sea (pineapples, bananas, mangoes, grapes, and so on) are usually picked underripe so they can survive longer in transit, and they tend to have extra fungicide applied for the same reason. . . . When they arrive at their destination, they are moved into ripening chambers, where ethylene gas is used to simulate a gigantic fruit bowl effect and ripen them.
>
> (Blythman, 1996:26)

In these computer controlled atmospheric chambers, fruit can be ripened at will, according to just-in-time schedules, and its shelf life extended up to five times the length of traditional storage. Since, however, this method of manipulating atmospheric conditions is considered a mere variation on the natural composition of air, governments world-wide have decided that there is no need for this kind of storage to be declared. My disquiet here is not so much with this particular method for simulating freshness as with the lack of public knowledge about it. I am concerned that those buying atmospherically controlled fruit and vegetables have no means of telling the difference between real and counterfeit freshness. In such cases, it seems to me, we are not dealing with a situation of informed choice but deceit: the vegetables are *meant* to look fresher than they are, supposed to simulate recent harvest and 'pure and untainted' condition.

Whilst the deceitful extension of the storage life of foods is not harmful in itself (as far as is known at present, that is), health hazards arise with the extension of food chains associated with industrial production. Since contamination of food may occur at any point between source and consumption, any lengthening of the food chain as well as any increase in the scale of the processes involved will increase the potential for health hazards and decrease the potential for tracing the sources of contamination. With food-borne diseases on the increase world-wide, we are faced with a vicious circle of ever more of the harmful kind of bacteria, fungi, moulds, protozoa and viruses in circulation in a context where expanding networks of infection and cross contamination are created through jet-setting, globally sourced foods. During the summer of 1996, Japan experienced its biggest outbreak of food poisoning to date which killed many people. In the UK, local and national newspapers abound in reports on food poisoning in hotels, restaurants, hospitals and care institutions. Each of the traded 'bugs' have their own unique temporal profile, that is, the typical period between ingestion and the onset of symptoms. The difficulty of dealing with this kind of public health hazard arises from the fact that the typical timespans can have a wide range of tolerance: where symptoms occur within a few hours, as in the case of botulism, the timespan between cause and effect may extend from seven to thirty days with listeriosis. (For a list of food-borne diseases, their symptoms and their temporal profiles, see Tansey and Worsley, 1995:61.)

The extension of our foods' storage and shelf life by means of chemical treatment, irradiation and genetic engineering adds further hazard potential to an already problematic situation. Since I am discussing hazards arising from synthetic chemical and gene manipulation elsewhere in the book (Chapters 1 and 7), I shall restrict myself here to comments about some of the temporal issues associated with the chemical treatment and the irradiation of food and some observations on additional areas of public concern.

Chemically extended shelf-life, for example, tends not to be declared. Consumers are not openly informed that their 'fresh' food has been chemically treated to simulate freshness over an extended timespan. The arguments to defend this secrecy are manifold: the chemicals are meant to have disappeared, which means they should have no traces by the time we get to eat the foods. They are meant to be harmless for humans. They are meant to be present in such small doses that they constitute no risk to human health. Thus, Joanna Blythman explains with reference to a range of post-harvest preservatives:

> Take potatoes. Even stored in the cold and dark, these would eventually sprout – the potato's way of indicating that it is beginning to deteriorate. The standard method of dealing with this is to treat them with Tecnazene, a toxic fungicide that inhibits sprouting. Because this is applied at what is deemed to be a 'safe' interval before sale, it does not need to be declared on the label. . . .Although these preservatives [Thiabendazole, Diphenyl, Oethophenylphenol and Sodium] have 'E' numbers (E230–3) which would have to be included in the ingredients listing in other food products, fruit and vegetables are mysteriously exempt from such labelling requirements.
>
> (Blythman, 1996:29)

Without declaration these chemical treatments remain invisible to the consumer. This means, we are unable to make informed choices about the food we are buying.

Consumers, we are told, are exposed to a minimum of risk because all the chemicals are thoroughly tested and, if applied correctly and with the appropriate time strategy, the chemicals are no longer in or on the fruit. Moreover, so the official argument goes, most natural foods are also toxic if taken to excess in terms of dose or frequency. Thus, rhubarb is toxic if we eat too much of it. Too much herbal tea can have a negative effect on health. Green potatoes contain solanin, a substance that is poisonous when taken during early pregnancy. Synthetic chemicals, so the promoters of such treatments argue, are no different: dose, frequency and timing are the issues that matter. As long as those who apply the chemicals use the correct dose at the right time and as long as the stipulated gap between application and consumption is rigorously observed, we are assured, preservatives are as safe as any natural food. But then, we need to ask, are all agricultural labourers literate? Are they all able to read and understand the instructions about the use of these chemicals? Can every one of them be relied upon to carry out the instructions to the letter? Do all businesses care about what happens to some anonymous consumer somewhere, some time, way down the trade chain? Is everyone along the way content to suspend the time = money principle and their habit of discounting the future for the sake of some unknown foreigners' safety? Can we be sure that no one will be tempted to increase their profit by speeding up the process, that is, by not waiting for the required quarantine period before offering the foods for sale?

Irradiation, the prolongation of the shelf-life of foods by means of radiation, had been hailed as the food industry's panacea for all its remaining ills associated with the long-term preservation of 'freshness'. It was to kill off all the bugs that had entered the food nets and increase further the time spans over which food could be kept 'fresh'. The irradiation of food works as a preservative in a dual way: first, it exposes food to high doses of radiation which achieve chemical changes that have a life-extending effect on the foods. Second, it kills some of the bacteria and organisms that are responsible for the decaying process. But, then, it also kills off vitamins, nutrients and positive bacteria. This answer to increasingly pernicious problems, therefore, soon came to be seen as part of the problem rather than the solution. In the UK, for example, some of the most eminent food scientists put their weight behind the Campaign Against Food Irradiation (CAFI) and were supported in their endeavours by organisations ranging from 'The National Federation of Meat Traders' to 'Parents for Safe Food' and the 'Women's Institute'. I shall deal with the safety debates that surround/ed this particular means of extending shelf-life in the next section. Here I just want to note that whilst irradiated food is required to be labelled by most countries, in the UK this stipulation does not apply for foods bought by restaurants and catering establishments and for processed foods where less than 25 per cent of ingredients have been irradiated. Again, the secrecy amounts to deceit since the people buying and eating those foods are none the wiser. With food that is sourced and traded globally, of course, the irradiation of food cannot be policed and is thus wide open to abuse.

At the end of this brief encounter with simulated freshness, we can see that the industrial storage of food has not only vastly extended the elasticity of the meaning of 'freshness' but, over the entire range of methods available to date, it has also blurred the boundaries between 'safety' and 'hazard': scientists, national governments and policy makers cannot come to agreements over safety and levels of risk associated with these issues. Scientific research produces contradictory results. Governments worldwide come up with opposing legislation. While businesses and governments are busy protecting the public's interests, health and safety by these means, members of industrialised countries have little opportunity to purchase genuinely fresh food at the point of ripeness. The production of freshly harvested food for immediate sale has been rendered uneconomical by the many ingenious ways to simulate freshness. Too costly, it has become the preserve of a select few, but even they cannot evade the web of deceit that has been spun on the behalf of citizens and in the 'interest of the public'.

### Safe until further notice: testing the poor

You can always pick a slug off your lettuce or rinse the leaves to remove greenfly but unfortunately, you cannot buy 'conventionally' produced food with an opt-out clause – the chemicals are an integral part of the deal.

(Blythman, 1996:278)

Safety, as I am arguing throughout this book, is difficult to establish in a context of time-distantiation and time lags, where damage and harm are being produced out of sight, below the surface, for often unknown periods of time and where the symptoms do not necessarily allow for a backwards reconstruction to originating sources and causes. Whether we are dealing with pesticides, post-harvest preservation treatments or irradiation, there are two opposing conventional responses to this situation: such processes and their effects are considered safe until proven harmful or dangerous until proven safe. Both, of course, are based on the false assumption that 'proof' is obtainable. Both are steeped in a Newtonian science understanding of the world that is principally inappropriate for time–space distantiated phenomena where the extent of time and space between initial condition and action and eventual symptom are irreducibly indeterminate. Moreover, not finding evidence of damage now does not mean that symptoms of irredeemable damage will not turn up tomorrow, the next year, in 20, 100 or 1,000 years' time.

We cannot prove the absence of a risk, . . . just as we cannot prove the absence of ghosts, say. No matter how many experiments you carry out where you find no ghosts, you cannot exclude the possibility that if you did one more experiment, you would find a ghost.

(Fritz Diel from the Food Preservation Institute, Karlsruhe, quoted in Elliott, 1990:45)

While this quote too suggests a bias towards a Newtonian science understanding through its use of the idea of proof, I like its powerful imagery: the notion that ghosts are there and not there and science has no means of establishing their non/existence. The opposing approaches to time–space distantiated hazards need, I want to suggest, to be rephrased into something like: we assume processes and methods to be safe because we cannot tell whether or not they are dangerous or, alternatively, we assume them to be dangerous because we cannot tell whether or not they are safe.

Despite assurances to the contrary, historical evidence suggests that governments the world over have tended to go for the first option: safe until proven harmful. Geoff Tansey and Tony Worsley (1995) cite examples of dietary changes, brought about by industrial mass production and preservation methods, that were subsequently found to have harmful effects on the population. In the UK, for example, it was retrospectively established that refined and canned foods had dramatically reduced the nation's health.

> The scale of the problem became clear during the First World War when the British carried out a mass medical examination of 2.5 million men in 1917–18. They found that 41 per cent of men supposedly in their prime were totally unfit for military service, mostly as a result of undernourishment.
>
> (Tansey and Worsley, 1995:45)

In Asia it took 15 years to establish the link between the refinement of rice and Beriberi, a debilitating disease that attacks the heart as well as the nervous and digestive system. The husks of rice, it was eventually realised, contained all the essential minerals and vitamins which were literally polished away with the production of white rice, leaving the population open to disease and declining health. Tom Elliott (1990:44) reports from India, a country that permits the irradiation of food, where a group of malnourished children were given chapatis from irradiated grain. The children were monitored and as they grew it was noticed that their cell growth, just like that of the mice that were tested simultaneously, showed abnormalities. Innocent until proven guilty, the programme was stopped only after the 'evidence' achieved a scientific level of proof. 'The history of agrichemicals', writes Joanna Blythman (1996:277), 'is littered with toxic substances that were considered safe and then consequently withdrawn when they were shown to be dangerous long after doubts had been raised about their safety'. The transformation of herbivores into carnivores and the insistence that it is 'safe' to eat meat from potentially infected animals – as long as their spinal cords and brains have been removed – is only the latest in a long string of 'safe until further notice' incidences associated with the industrial way of life in general and its production of food in particular.

What also seems clear from the historical data is that the poor have been elected to be the canaries of the food system. By ensuring that new industrial alternatives to traditional foods are cheaper, the promoters of these foods are guaranteed a captive consumer group, keen to try the latest innovation. Tinned meat and vegetables, refined foods, jetsetting foods from across the world that have been chemically treated for beauty and storage, irradiated food, hormonally treated milk, baby milk contaminated with Chernobyl radiation – all are foods that have been duly tested by the poor, with and without their knowledge and/or consent. Choice does not even come into it when the cheapest option is the *only* alternative to starvation.

## Future prospects: reflections on the scope for sustainable action and change

Given the gloomy prospects, what scope is there for alternatives? What potential is there for change towards safer food? How might farming be reinstated in its role as provider of healthy nourishment? What openings might there be for the protection of people present and not yet born? In response to the issues raised in this chapter I would like to identify some points of departure from established socio-political and socio-economic traditions and make some suggestions for change with reference to political action, consumer rights, retail practices and farming praxis.

### 'Back to basics': local and seasonal food as right

The British government under John Major came up with the slogan 'Back to Basics', a poorly thought out idea that did the Tory Party more harm than good. Applied to the food industry, however, this political catch-phrase could be an eminently sensible motto and ideal for achieving a more sustainable food production on the one hand and a dramatic reduction in public health hazards on the other. In a context of globally sourced food where decontextualisation and a-seasonality are centrally implicated in the production of food-based health hazards, 'Back to Basics' would mean the facilitation of locally grown, seasonal foods. Appropriated for food production, such an emphasis on basic principles could reduce not only the need for chemicals and transport, but the cost of the national health service. I want to suggest that *locally grown, seasonal, genuinely fresh and non-processed food ought to become a basic citizen right which supermarkets and other retail outlets are legally bound to supply*. This would not deny anybody the right to buy the jet-setting food they have become accustomed to. Globally sourced, artificially preserved food would simply be offered alongside the basic, unadulterated, ripe local food. For consumers it is important to retain the capacity to choose. An eco-dictatorship that outlawed jet-setting foods would achieve nothing but resentment. What would be essential, however, would be to stop the current pricing magic whereby the costs for such foods are spatially and temporally externalised. Instead, the full costs in terms of energy, health and environmental damage would need to be internalised and reflected in the price of foods.

In order that people may exercise their right to local and seasonal foods, I would like to propose the establishment of a star system by which degrees of locality are identified: food produced within a 10-mile radius, for example, would be awarded 5 stars; food from the local county, district or parish would qualify for 4 stars; from the region 3 stars; from the country/state 2 stars; from the continent 1 star; and anything from further afield would get a 0 star rating. A similar principle could be applied to natural freshness, degree of processing, and level of chemical intervention at all stages of the particular food's history. This system would be simple to put into place, boost local economies, and dramatically improve food safety for all citizens. Moreover, it would do no harm beyond affecting the economic prowess of those who are currently either busy promoting or actively engaged in practices that create long-term potential and actual hazards.

### Changing economic practice: comprehensive food histories as consumer right

Many countries of the industrialised North are currently experiencing a 'retailing revolution'. Competition between retailers is no longer merely about prices but about 'customer service'. Citizens concerned with food safety need to make sure that the next retailing development is about the provision of information, that competition between companies will be about who is providing the most comprehensive information for their customers; who offers the best stories about the life-history of their meats, vegetables, pulses and exotic foods. When the awareness that 'we are what we eat' comes together with such information in a context of equitable prices, then we can talk of an economic climate of consumer choice. Full information on foods, we need to appreciate further, would mean the entire history including the gory details: 'this apple has been subjected to irradiation and treatment with toxic herbicides, pesticides and fungicides'; 'when still alive this pig had been given growth hormones, was treated with valium just before slaughter, and had been administered weekly doses of antibiotics from birth'; 'on the basis of knowledge to date this can affect you in the following way . . . '. Of equal interest to the consumer would be information about the way: a crop had been grown, the eggs had been produced, the particular food-animal had been kept over the course of its lifetime. Whenever the right to such comprehensive information is in jeopardy, it is not the right that should be sacrificed but the particular offending food that should be eliminated. That is to say, only foods for which a full and traceable history is available should find their way on to the shelves of shops and supermarkets. Putting into place a globally policed food information system of this kind would go a long way towards preventing the current practice of rending invisible the chemical treatment, irradiation and genetic engineering of foods.

The development of this system, moreover, would create a whole raft of new and interesting jobs to research, track and write about the individual food histories: job creation for the noble purpose of ensuring that customers know exactly what they are about to buy and internalise. Such a system may be 'inconvenient' and cumbersome in the beginning but once established it will be merely a question of attitude: where there is a will there is a way. The new system would not only be a logical extension of the 'service revolution' but also, like previous socio-economic innovations, work to the political and economic advantage of those who initiate and promote its revolutionary principle. The right to locally produced food and comprehensive food histories will of course require appropriate laws to be passed that are valid globally. There can be no second-class citizens when it comes to these basic human rights for the twenty-first century.

### Tempering agricultural pace: owning time and the means of reproduction as farmers' rights

For farmers there arises the need to re-appropriate the means of reproduction. If farming is once more to be associated with health-giving nourishment and sustainable practices, farmers need to re-establish their basic right to own and control their seeds and semen, their ancient crop varieties and their full range of locally evolved and adapted breeds. This will return to them as a right the construction of the long-term future, the capacity to plan and think in generations. In conjunction with the human right to seasonal and locally grown food, the farmers' right to ownership of the means of reproduction would allow farmers once more to attend to their primary task of growing food for people. Where currently farmers spend much of their time on EU bureaucracy, they will need to shift their attention to a bureaucracy of a more meaningful kind: keeping meticulous life histories of their animals and crops. It would enable them to transcend the current situation of growing for export, EU storehouses, and incineration. It would help to bring an end to the subsidised trading practices that render local produce uneconomical. Thus, it would stop the current situation where local farmers have to let their fruit rot on trees and plants, leave crops in the fields and pour milk down the drain. Compulsion, be it for or against export and global trade, however, should not be part of the scheme. If farmers want to grow for export, fine. What is important is the fair treatment and positive revaluation of local, seasonal, freshly harvested, unadulterated produce in relation to globally sourced foods of simulated freshness that fundamentally depend on external energy imput for transport and preservation.

To achieve this, farmers need not only to re-appropriate the ownership of reproduction but also of time, since the latter is a central precondition to sustainable agriculture and the production of wholesome food. It is crucial in a number of ways and for a number of reasons:

- To acknowledge the here and now in its rhythmic and seasonal particularity is to know the future as patterned creativity. It allows the producers of food to prepare for in/determinate repetition of unique futures and to think not in the severely limiting time frame of economics but the long-term time-scales of natural re/production;

- To recognise the importance of locality and seasonality is intimately tied to an appreciation of the species-specific *Eigenzeiten* of animals, plants and ecosystems: not the abstract universalised and rationalised time schedules of non-stop industrial production, but the rhythmic uniqueness of species-environment time which provides the appropriate guides for agricultural practice and the production of high quality foods. Ownership of time means farmers (should they wish to) can structure the temporal organisation of their lives to suit their animals and their crops, their own lives and their long-term plans. Where the choice is for the rhythmicity of nature and the cosmos, the pace of agricultural life will be tempered and more measured;

- Ownership of time with its potential for reduced pace and extended timeframes of planning and concern may entice farmers to continue doing this important job and to think once more in terms of generations. For the consumers of the produce it translates directly into food quality not in terms of beauty and uniformity but taste, freshness and nutritional value. For societies it will reduce the collective stress level, boost local economies and improve the health of their members.

There is nothing to lose but a web of unsustainable practices that benefit an exceedingly small proportion of the earth's people and potentially harm humans and other species together with their socio-physical environment for an open-ended future and unspecifiable numbers of generations.

*Plate 5* Spreading plastic on strawberry farm by William Garnett. *Source*: the photographer.

# 5

# MEDIATED KNOWLEDGE
## Of time-lags and amnesia – reporting on BSE

## Introduction

On 20 March 1996, in a televised statement to the House of Commons, Stephen Dorrell, the then Health Secretary, announced that ten cases of a new form of Creutzfeldt-Jacob Disease (CJD) had come to light and that the Government's Spongiform Encephalopathy Advisory Committee (SEAC) 'has concluded that the most likely explanation at present is that these cases are linked to exposure to BSE before the introduction of the specified Bovine Spongiform Encephalopathy offal ban in 1989'. This announcement has set in motion one of the most intense and protracted media coverages of any UK public health issue since World War II. No other socio-environmental hazard, threat to public safety, environmental degradation or pollution, not even the radiation fall-out from Chernobyl has attracted a similar level of attention from the British media.

Bovine Spongiform Encephalopathy (BSE) in UK cattle has been a media issue since it was officially identified in 1986. From then on it has periodically dominated the news and been subject to investigative television programmes such as *Dispatches*, *World in Action* and *Panorama*. The public announcement, in March 1996, however, marks a watershed, negating ten years of political and scientific assurances about cattle being the 'dead-end hosts' of this disease. From then on, something that has arisen in the media as an occasional if persistent scare took on epic proportions as an invisible threat. During the first week of reporting on the 'crisis', every single newspaper in the UK gave it extensive attention, offering coverage from every possible angle. Thus, on the following day, the *Independent*, with fourteen separate articles on the subject, was no exception among the broadsheet press. It seemed as if every correspondent who could possibly have something to say on the matter was told to come up with their perspective on the issue. Even though there were unambiguous attempts to define the BSE–CJD connection as yet another food scare – of which Britain had extensive recent experience with listeria and salmonella – this health hazard turned out to be more persistent and troublesome than any of the others. Since then and up to the point of publication of this book, BSE and to a much lesser extent CJD had remained a staple ingredient of news, with significant flare-ups of intensification each time the drip-feed system of disaster announcements added new information that contradicted previous pronouncements and assurances of both a scientific and political kind.

The news media's role in this story is a complex one. First, we cannot even begin to guess how the entire fiasco would have developed if the *Daily Mirror* had not leaked the story. Second, the media both reflect the politico-economic *status quo* and question it; they report on issues and simultaneously raise public consciousness about them. Third, they are informant, analyst, mediator and creator of a reality that is mostly inaccessible to a large proportion of their readership and audience. At one level, therefore, the news media are a mere channel of information, at another they are defining the parameters of the issues and at a further level still they are the constructors of knowledge. Through their mediation, interpretation and translation of otherwise inaccessible knowledge into publicly accessible form, newsworkers are not only prime sources of public information but also the principle social theorists of contemporary industrial societies. As such, they carry a heavy burden on their shoulders, a responsibility that does not sit comfortably with their own self-perception, that is, their understanding of themselves as harbingers of news, disseminators of matters of human interest and providers of a critical perspective on the more shady aspects of socio-political and socio-economic life. Socio-environmental hazards, I argue in this chapter, require of the news media a level of social analysis they are at present poorly equipped to provide. That is to say, it is not just scientists, economists and politicians who are struggling to come to terms with the new context and the environmental problems that have arisen from the industrial way of life, but the media too are facing issues and dilemmas for which their traditions are proving inadequate. Again, time is centrally implicated in the difficulty.

In this chapter I therefore illuminate from a timescape perspective the central role of the news media as source of information and social analysis about environmental hazards. Focus on the BSE–CJD hazard brings to attention both the potential and limit of this media's capacity for that extended public role. Where else would citizens get their analysis and commentary about BSE, hormones in meat, and the latest developments in genetic engineering? Who but journalists would interpret for them scientific findings, academic papers, and the bureaucratic jargon of national and international policy? In a context where neither government nor business is trusted, who could citizens turn to for interpretation and analysis of all the debates and contradictory findings? Who beyond the media would champion their cause in a way that is accessible and understandable, speak their language, express their concerns? Not academic social science, I am afraid, and yet, media theory falls far short of what academic theory has to offer. This situation, therefore, leads me to consider what might constitute socially responsible, social theory for mass consumption. The chapter ends with some reflections on this matter.

## Timescapes of the media: reporting environmental hazards

### *The context of scientific uncertainty*

A substantial number of socio-environmental hazards arising from the globalised industrial way of life, as I argue throughout this book, tend to be invisible, inaccessible to the senses until they materialise as symptoms after indeterminate periods of latency. This applies whether the threat is radiation from nuclear power or the damaged ozone layer, genetically modified organisms which have been released into the environment or residues of pesticides in the food, oil pollution below the surface of the sea or the contamination of UK dairy herds with BSE. Once symptoms have arisen in the form of cancers, deformities, allergies and contamination of marine life or meat, for example, it is exceedingly difficult and sometimes impossible to establish verifiable connections between those symptoms and their causes. During the invisible periods of latency the provable links have been broken, causal connections severed. This discontinuity has implications for the scientific definition of a problem, the understanding of causes, and the formulation of potential solutions. All of which, in turn, influence the way the media engage with these issues.

Whilst scientists are fully aware of the complexity of the situation and the severe limitations for unambiguous and fast resolutions to previously unknown and time–space distantiated socio-environmental hazards, politicians and journalists on behalf of the public demand proof and certainty. On the basis of their Newtonian habits of mind and physicalist assumptions they clamour for science to provide proof of (physical) causal links between, for example, cancers and nuclear power plants or Scrapie in sheep, BSE in cattle and CJD in humans. Scientists, of course, are unable to supply such incontrovertible evidence, establish such absolute connections. Consequently, they are exasperated when neither politicians nor the media take seriously their message that such certainty and proof is not to be had, that it is unobtainable for phenomena where unknown and unknowable factors far outweigh what has been established and is known. They find it frustrating to witness the transformation into the language of certainty and fact what they had expressed with utmost caution in terms of hypotheses, conjectures and probability. Equally disconcerting for them is the pressure for the dangers to be quantified, to be expressed in terms of risk, and for cost to be calculated in purely material and monetary terms.

In a context of great social dependence on scientific 'solutions', it seems, admission of ignorance is not regarded by the media, the politicians and the public to be an appropriate or acceptable scientific response. Scientific uncertainty is considered an anachronism. The discrepancy between expectation and changed context of scientific knowledge creates dissonance and widespread indignation. For over two hundred years, science (of the traditional, Newtonian kind) had been the trusted and revered provider of certainty and truth, of unambiguous and unassailable knowledge. Now that this knowledge is thought to be desperately needed to solve contemporary societies' socio-environmental problems (for whose creation science has to carry most of the responsibility) the scientists' caution and language of uncertainty is interpreted as a 'fudge' by much of the broadsheets and treated with derision by most of the tabloid press. Where scientists are resistant to come up with quantifications and facts, journalists and politicians tend to do their job for them: the vague and propositional language of science gets translated into politically and economically acceptable certainties and assurances. Whatever stands in the way of such corrections and whatever cannot be quantified or established with certainty is simply left out. This has the effect that irrespective of the high level of uncertainty and indeterminacy that surrounds socio-environmental hazards such as BSE, their representation in newspapers and news broadcasts are brimming with economic facts, figures and projections, with reports about what politicians did and what they said to whom, with analysis/speculation about who has been and/or will be affected when by what actions and, finally, with statements about who is banning beef from where on what basis. What is missing from this mass of factual writing and reporting on the subject is a frank discussion about all that is uncertain and the effects of this inescapable indeterminacy for polit- ical action and the potential for securing public safety. The crucially import- ant *Wirkwelt* of the hazards is silenced into oblivion, creating yet another velvet void, this time at the centre of media hyper-activity.

It is, however, not just the way the news media handle this context of uncertainty that brings about changes in public perception. Where the news media act as means or channels for information, scientists and politicians are in a position to communicate directly with a mass public. Whilst science normally tends to be presented on television in highly structured and man- aged contexts – in prestigious television programmes such as *Horizon* or *QED,* for example – the BSE–CJD crisis put scientists in unrehearsed and rather less dignified circumstances of communication. It is my reading of the situation that the public standing of science has been badly damaged by the BSE–CJD episode. Science has suffered a blow to its respectability and its credibility not only through mis/representation by the press but through self-inflicted disastrous 'performances' on television. As the crisis broke, for example, viewers were able to watch Professor John Pattison, the chairman of the Government's BSE advisory committee, change his story from one day to the next: on Wednesday 20 March,1996 he stated that the potential number of humans contracting CJD could be anywhere between two-figure numbers and 500,000, thus indicating that the *potential* was there for a major epidemic. Either scenario, he suggested, was possible; science at this point simply did not have an adequate basis upon which to make predictions or calculate risks. The very next day, however, this same scientist insisted that due to the regulations which the UK Government had put in place in 1989, the risk of contracting CJD was very, very small indeed. On the basis of these measures, he was adamant, the public could be assured that, 'in the normal sense of the word', British beef was 'safe' to eat. Surely, the data surrounding these regulations are no more certain than the causal connections between BSE and CJD. Given that they had not been enforced by law, how could John Pattison be certain that they have been strictly adhered to – which they have not? How could he possibly know that no cross-contamination has taken place between feeds, between carcasses, between animals, between species – which it has? How could he be sure that no illicit trade has emerged around the banned offal – which it has? This list of uncertainties could easily be extended to equal that associated with the causal connection between BSE and CJD. The people I had talked to about this extraordinary turnabout were convinced that John Pattison was told by the Government what to say. On the basis of this bizarre overnight transformation, and many equally stunning somersaults, science came to be seen as nothing more than a puppet attached to the Government's purse strings and dancing to its paymaster's tune. What credibility science had before – even if the public's expectations were unrealistic – came to be thoroughly discredited over the BSE fiasco.

The UK Government, of course, comes off worse still. Whether on television, the radio or in the press, whether through statements in interviews or through reports, the business of government was transformed into a farce. Almost day by day the British public was told a different story, promises were made and broken, assurances were given and retracted, farmers were given conflicting messages, EU partners were left speechless. Even staunch supporters of the Tory Government admit that the BSE crisis was and still is handled very, very badly. Less complimentary comments included 'a bunch of headless chickens', 'incompetent talking heads', 'hyperactive delinquents', 'exclusively concerned to please their paymasters', 'interested only in saving their own skins'. Since British society does not suffer from collective amnesia, Stephen Dorrell's fateful announcement was received in a context of eight years of governmental assurances that British beef was safe. Thus, Douglas Hogg, Agriculture Minister at the time of the announcement, had consistently insisted that British beef could be eaten with confidence. John Gummer, his predecessor, had 'not a scintilla of doubt' that British beef was not only safe but the best in the world and demonstrated that faith to a television audience by feeding his daughter Cordelia a beef burger. 'Ministers', as Steve Connor and Michael Prescott remind readers of the *Sunday Times*, 'went on repeating that beef was "perfectly safe": that eating it carried "no conceivable risk" ' (ST 24/3/96:12). Mr Dorrell's statement, moreover, has to be appreciated alongside a widely publicised letter by John Major, the then Prime Minister, to the mother of a young person who died from CJD in November 1995 in which he stated categorically: 'I should make it clear that humans DO NOT get mad cow disease' (*Independent* 21/3/96:3, *Daily Express* 21/3/96:5). This series of false assurances, contradictions and inappropriate responses and the inability of both politicians and scientists to take charge of and deal with BSE and similar threats has led to widespread public disillusionment. Robin Grove-White (1996:270) identifies a 'mounting fatalism and cynicism' and a loss of faith in the capability of formerly trusted institutions to protect the public from the in/visible and im/material harm and dangers associated with the contemporary, industrial way of life.

The way the news media deal with such issues, therefore, is crucial to the public's perception of and response to such socio-environmental hazards. In the case of BSE and CJD, for example, it makes a huge difference whether or not the story is told purely from a UK perspective, from the position of the Government or the afflicted, from a sympathetic or antipathetic scientific stance, from a concern about health or economics. In the face of socio-environmental hazards, the media are literally what the name indicates – mediators of otherwise inaccessible knowledge and information beyond the realm of everyday experience. This mediation involves the translation of scientific knowledge into an accessible form, explanation of political action, presentation of relevant data, and the interpretation of potential and actual social effects. Clearly, the story to be told will differ from newspaper to newspaper and between television companies. Along the same lines, the news media's perspective on the credibility of government and science depends very much on where their political loyalties are located.

And yet, despite widely differing social perspectives, there are discernible shared patterns and limitations across the breadth of the news media. These common features relate to basic assumptions and, arising from these, to what kind of stories are being told. Like most of the society within which they are functioning, UK journalists tend to be steeped in the habits of thought that permeate the knowledge they mediate. In other words, journalists are predominantly working with a linear perspective, Newtonian science assumptions and an economic frame of reference. Furthermore, the news media's difficulty of appropriate engagement with environmental hazards in general, and the BSE–CJD link in particular, extends beyond such Newtonian habits of mind to the very definition of their profession with reference to both the subject matter and structural features of journalistic practice. It is therefore important to bring to your attention some of the temporal features of these two aspects of media production by focusing specifically on newswork.

## The timescape of news

News reporters tend to differentiate between two kinds of news, both with distinct temporal characteristics. The first kind are 'hard news' which encompass the factual and temporally bounded realm of accidents, crimes, disasters, conflicts, one-off events, statements, and outcomes of actions that are newsworthy today. These cover issues of public interest and concern that are pertinent here and now. Such now–here phenomena tend to be perishable. They have a limited life as 'news' since, as Allan Bell (1995:306) points out, 'the next edition renders them obsolete'. Such hard news items operate within the replacement cycle of hourly news bulletins, fixed daily news reports and papers published on a daily and weekly basis. 'Soft news' in contrast, are phenomena with a longer shelf-life, issues of public concern for which it does not matter whether they are reported on today or the week after. Environmental hazards tend to fall in this latter category. Given that most of them are of a long-term nature, the specific time and day chosen for their discussion is not that important. Irrespective of their degree of importance, matters concerning the environment tend not to 'pop up' today and be gone tomorrow. Instead, their social pertinence is long-term, chronic and often cumulative. This gives them an air of relatively stable and predictable continuity, punctuated only by political actions, new research findings and the emergence of unexpected symptoms arising somewhere, some time. While hard news require almost real-time reporting – the faster the better and instantaneity is best – environmental matters, from ozone depletion to pesticide contamination of soil and water, do not command the same sense of urgency. Instead, during their 'stable' phases, they are seen by the news media of television and press as useful filler items for the days when there is nothing more urgent, timely or pertinent to report.

'News' is traditionally associated with specific characteristics, so-called news values, which would include novelty, timeliness, recency, immediacy and urgency. These are clearly descriptions that fit accidents, crimes, disasters, and political events. (See Bell, 1995 for an extended discussion on this topic.) They rarely apply to globally dispersed and time–space distantiated environmental phenomena that reside nowhere in particular and go on for years without resolution. Long-term, chronic hazards do not allow journalists to limit their story to the here and now. Instead, they require historical contextualisation, since without such wider historical background the information may be meaningless. Such matters of social concern are likely to demand frequent coverage which, in turn, needs to be carefully planned with respect to what angle to take when and which aspects to cover when, so as to avoid telling the same story many times over. This kind of approach to 'news' is almost a contradiction in terms. The long-term continuous pertinence of environmental hazards, therefore, constitutes a major challenge to newswork. News as the delimited here and now of events has to be rethought in the context of chronic environmental hazards associated with the industrial way of life.

At the practical level too, those hazards pose significant difficulty for the tradition. Their complexity frustrates the need to be fast and succinct; it hampers the need to put together the stories at great speed and then to present them in predefined spaces or sound-bites. Moreover, complex environmental 'news' necessitate a shift in emphasis from description to analysis, from accident-oriented information gathering and reporting of isolated facts to interpretation, explanation and analysis of conflicting data, ambiguities and unknowns. Environmental hazards, by default, turn news reporters into social theorists. The question is, how well have newsworkers risen to that challenge and to this new role? Newspapers and newscasts are not at all well suited to that new demand. The speed of turnaround, the competition between papers and channels, and the need to sell maximum copy or have maximum listening/viewing figures do not bode well for environmental hazards getting the treatment and attention they would need. As one American journalist said at a recent conference (*Reporting the Environment*, Cardiff, 19–21 May 1996): 'our prime task is to sell newspapers and the environment is not a sexy topic'.

Bearing in mind these obstacles I have identified, how do environmental hazards fit the requirement of 'sexy' or emotive topic and mediagenicy? The remoteness in time and space of such matters as global warming, ozone depletion, or the global permeation of hormone-disrupting chemicals is difficult to present with the necessary urgency and emotive immediacy to make a good newspaper-selling story. Moreover, most of these hazards do not lend themselves readily for representation by visual images that symbolise the issue and elicit deep-seated fears. Oiled birds and stumbling cows, in contrast, constitute differing degrees of such symbolic expression. This makes an oil tanker accident and a public health issue such as BSE more mediagenic than hazards that are less visual and temporally more remote. The temporal, spatial and visual features of the BSE–CJD link are therefore worthy of further attention.

### The BSE–CJD link: a new kind of hybrid news

While the BSE hazard, over its ten-year history, belonged predominantly to the category of 'soft news', once the UK Government announced the likely link between BSE and CJD, a most horrific deadly disease, the issue was catapulted into the 'hard news' category: it became an urgent now–here phenomenon that touched profound public fears and had the potential to destroy livelihoods. As such, it required and commanded immediate and extensive media attention. The announcement that BSE-infected meat is a likely cause of CJD, hit Britons and the citizens of continental Europe like a bombshell. Everybody recognised themselves as implicated: meat eaters and their children; vegetarians who were told that foods such as cheese, fromage frais, and fruit jelly contain beef products; people who had undergone growth hormone treatment during childhood. Moreover, while everyone clearly had choices to make about the future consumption of beef, the biggest risk relating to CJD was located in past actions, thus outside today's parameter of choice and control. Harry Thompson (former producer of the television programme *Have I Got News For You*) expresses this no-win situation when he writes,

> I don't actually want to take a risk, thank you, whether they turn out to be minimal or not. There is a 'minimal risk' involved in putting your head in a blender during a power cut, but no one in their right minds would do that either. . . .The trouble is, it's probably too late.

The beastly little prions are already in there, skulking in the grey matter, waiting to pounce.

(*Guardian* 31/3/96:3)

The people whose economic livelihoods are threatened or destroyed had no choice either. In order to improve milk yield, for example, farmers supplemented grass and silage with protein-rich feeds. Little did they know that this transformed their herbivores not only into carnivores but cannibals. The information on the feed bag did not specify that the listed protein was not of plant but animal origin let alone that it contained offal and bones from cows, pigs and scrapie infected sheep, that it included sawdust, feathers and faeces from chicken farms and other similarly extraordinary materials. This unspecified protein, the public is told, is responsible for devastating large parts of the rural economy of entire countries, farming communities and the meat industry, transport and export businesses, wholesaling and retailing.

At this level, therefore, one can easily see how the announcement of the possible BSE–CJD link corresponds with news values of now–here recency, immediacy and timeliness, of the unexpected, the shocking, and the socially pertinent. Its emotiveness is unquestioned. Its dramatic content and the potential for personal disaster is enormous. Pictures of staggering cows, empty cattle auctions and young people reduced to a vegetative state are emotive symbols that express not just people's fears but the conflicts of interest that permeate the manifold outcomes of this socio-environmental hazard. The mediagenic credentials are assured. With Stephen Dorrell's announcement the long-term, periodically newsworthy story has been transformed into a lead story that was to stay at the top of the news agenda for ten days and refused to fade into the background for the months that followed. As such, it received the 'treatment' metreed out to any major news story: it got coverage from all possible angles ranging from human interest to economics and science, from personal accounts to macroeconomic data.

At another level, however, this socio-environmental hazard does not conform to the 'ideal type' of news: it is not a once-off issue. It is not an event that is bounded in time and space. After seventy-five years of CJD and ten years of BSE, this 'new' threat to public health is not all that novel and unexpected either. Unlike the quantifiable facts associated with more conventional hard news, there is little about the BSE–CJD link and the two diseases' respective causes that would count as incontrovertible fact. Instead, this hazard is shrouded in mystery, permeated by uncertainty and marked by scientific disagreements. Its 'facts' are largely unknown and it is consequently open to almost limitless hypotheses, conjectures and speculations. On the one hand, it looks as if this hazard has successfully eluded political news management since it could not possibly have been the Government's intentions to show themselves in the worst possible light. On the other hand, however, it could well be due to the influence of spin doctors (the people responsible for managing news on behalf of their political bosses) that the real issues have not been discussed, questions of responsibility have not been posed, legal actions (if there have been any) have not received high profile media attention. Moreover, the story seems destined to 'roll on' even though the media are periodically showing signs of fatigue and burnout, eager to move on to the more familiar ground of less intractable news and keen to let this public health issue fade from the limelight back to the more manageable sphere of soft news. Finally, both BSE and CJD have a long history that forms an essential part of the present. Without this history, the significance of the present coupling cannot be understood. Despite the public's need for knowing the histories of those two diseases as well as the agricultural, political and medical actions associated with their respective developments, the public presentation of historical information is generally not considered an integral part of news. News, as I suggested above, is presented succinctly and factually as clearly delimited now–here phenomena. The history of a phenomenon thus tends not to form part of its news value. This clearly constitutes a difficult situation for the 'reporting' of such environmental hazards.

By 21 March, the day after the Government's announcement, the BSE–CJD story takes up several pages of every UK newspaper and is the lead item of every news broadcast in the land. Overnight a long-standing problem is transformed into an inter/national disaster story with almost infinite angles: political, scientific, economic, legal, human interest, health, nutrition, to name just some of the most obvious ones. Over the next months, up to the point of this book going to press, the media would move through all of these angles. Some papers, such as the *Independent* (22/3/96:1,4,5,12,13), covering virtually all of them on one day, others moving from one perspective to another over a protracted period. On some days a committed reader could find the full range of foci by reading horizontally across the breadth of the newspapers' coverage and listening to the news programmes of radio and television. Despite this massive and intense media interest, however, actual information was not that easy to extract from the thousands of pages of newsprint on the topic. Given the complexity of the issues on the one hand and the time constraints associated with news reporting on the other this should not surprise us, nor should the realisation that the hazard itself remained curiously disattended – an invisible, unspoken, ghostly shadow. This means, the media were and still are exclusively concerned with the *Merkwelt* of BSE and CJD, that is, with the outcomes of their invisible time-distantiated processes and actions. The *Wirkwelt*, that which gives rise to the symptoms, in contrast, is largely neglected and thus negated. In a context of news reporting it could hardly be otherwise.

The complex temporal issues entailed within the BSE–CJD story were and still are difficult to embrace within the context of news reporting. I have already mentioned that the history of these two converging diseases had not been given much attention. Some background information was provided but this, as I show below, was scattered among hundreds of articles in different papers and therefore not easily accessed and appreciated. The *Daily Mirror*'s (27/3/96) timeline which charts the sequences of events from first identification of a problem to the EU ban is an exception to the rule. Similarly neglected has been the important issue of the long and vastly variable incubation periods of the two diseases with all their implications for actions. Moreover, no one thought to consider in depth the effect of the politicians' uncertainty, hesitancy, and repeated change of mind in temporal terms. Instead the papers were full of calculations about lost export revenues, the impact on GNP, and the costs to farmers and retailers, the rendering and slaughter industry. An extraordinary amount of attention has been given to nationalistic posturing and the wheeling and dealing between national and European politicians and scientists. Attention on measurable and calculable physicality, events, actions and statements reformed this largely unsuitable news topic into something that could be managed within the high pressure world of instantaneous news production.

Where then does this adaptation of a health hazard to the professional, cultural and structural constraints of newswork leave the concerned citizen for whom the news media are the only source of information? To illuminate the substantial gap between media and public concern, I first place myself in the role of implicated citizen before I reflect on this situation in the later sections of this chapter.

## Reading about BSE and CJD: the first four days of news

Like the people around me, I lived and experienced the news about this socio-environmental hazard as an implicated citizen: speechless, concerned, disgusted, cynical, worried, but, above all, desperate for information. Of course, BSE was one of the many invisible hazards I was focusing on in my research but when it became 'real' to the extent that it was played out as a 'live drama' on a daily basis, affecting my family, myself and the livelihoods of the people around me, it seemed more important than ever to look at the issues from the embedded, embodied, personal, now–here position of implicated participant. Not the university's research facilities, therefore, but newspapers and broadcasts were to provide the bulk of my information. The coverage in the press, on radio and on television was supplemented, after a lengthy delay, by papers from the Ministry of Agriculture, Fisheries and Food (MAFF) which informed the farming community about new regulations and bureaucratic stipulations. The newspapers I consulted included the *Daily Express* (DE), *Daily Mail* (DM), *Daily Mirror* (DMi), *Daily Telegraph* (DT), *Farmers Guardian* (FG), *Financial Times* (FT), *Guardian* (GU), *Observer* (OB), *Independent* (IN), *Sun* (SU), *Sunday Times* (ST), *South Wales Echo* (SWE), *The Times* (TI), *Wales on Sunday* (WS) and the *Western Mail* (WM).

As implicated citizen, I experienced an initial flurry and frenzy of media activity in which concerns over health matters rapidly gave way to economic considerations and political concerns of a nationalistic nature. So swift was the shift of focus from concern about public safety to anxiety about job losses and a tarnished national image that neither the scale of the problem nor analysis of its causes got the urgent and up-front attention that I thought they required: critical reflection on what had gone wrong and consideration of how one might change the system in order to avoid similar disasters in the future were effectively sidelined. Like everyone else, I could not fail to notice the almost obsessive preoccupation with the EU and European responses to what was after all an appalling UK situation that never should have happened. It became unpatriotic to talk about anything being wrong with British beef, British farming, the British industrial way of life. I had experienced that mood before. It was rife during the Falklands 'crisis'. In the face of potential outsider criticism, it seems, the country pulls together and blames the rest of the world, but preferably continental Europe. And yet, within this huge wave of nationalism and economic concern about the beef industry and the country's export markets, there were critical voices offering alternative interpretations of the crisis and its solutions.

For me, the most striking thing about the first few days of news on the potential BSE–CJD link was how difficult it was to extract information from the immense amount of newspaper space allocated to the subject: the profusion of news stories, written from every conceivable angle, was matched only by the dearth of actual information of the kind that would have been helpful to parents and the vast number of people whose economic livelihoods were severely threatened. An article in the *Independent* by Charles Arthur illustrates my point. Entitled 'To Eat or not to Eat – the Facts behind the Disease' this piece posed some key questions and gave answers to each of them. The following extract illustrates the level of non-information that was prevalent during the hype of those early days after the public statement.

> How and when did it [BSE] start?
> Scientists are still uncertain whether it started from feeding cattle with the remains of sheep infected with 'scrapie' (the equivalent to BSE), or if it arose spontaneously in cows.
> How much British beef is infected?
> The number of cattle with BSE has fallen since hitting a peak in 1992, but some scientists estimate that thousands of cattle in the early stages of the disease slip through checks and enter the food chain.
>
> (IN 21/3/96:2)

I am sure I am not the only one that was left speechless in the face of this level of ignorance and this mixture of non-information and manipulative incitement of fear. Significantly, it was not until eight days after Stephen Dorrell's announcement that I found figures in the *Daily Express* (27/3/96: 13) which made clear the extent of the BSE epidemic in the UK and presented in accessible form the difference in scale between cattle affected in the UK and those of other European countries: UK 161,663, Switzerland 205, Republic of Ireland 123, France 13, Portugal 3, Germany 4 – the rest of the cited countries had figures below that and each of these cases were cows imported from the UK. (Once I had collated figures from articles across the breadth of the papers I found that while the detailed figures did not tally between the papers, and even between articles within papers, there seemed to be an overall agreement about the order of magnitude and difference between the UK and other countries.) The scale of difference certainly puts into perspective any talk about other countries having the disease as well. By the time I read these figures, however, a very different story had been forming in my mind. This means that even an avid reader of papers could have been left with the impression that BSE was not a primarily UK-based disaster. Such misperception was fostered by widespread insinuations – especially but not exclusively by the tabloid press – about other countries also having BSE but being less forthcoming than Britain with their public acknowledgement of that fact. The late timing in accessible form of this rather crucial piece of information, therefore, was significant for public interpretation of the issues. Given that the media were the *only* source of public information during a situation that could rightfully be classified as a national disaster, the press in particular did not serve the public very much better than the government responsible for safeguarding its citizens' safety and economic livelihood.

In contrast to the coyness about providing figures that would indicate the scale of the UK problem, the Government's new regulations were widely reported. These had been announced on 20 March 1996, alongside Stephen Dorrell's fateful disclosure to Parliament that ten young people had died from a new strain of CJD which meant that a possible link between BSE and CJD could no longer be discounted. The new measures are interesting in as much as they give insight into what the UK Government considers to be the root of the problem. That is to say, in the face of a potential epidemic in humans that could in 10–30 years' time replicate the disease pattern in cattle, the Government announce the following new regulations:

1 Carcasses from cattle aged over 30 months must be deboned in specially licensed plants supervised by the Meat Hygiene Service and the trimmings kept out of any food chain.

(IN 21/3/96:3, FG 22/3/96:1)

2 The use of mammalian meat and bone meal in feed for all farm animals {is to} be banned.

(IN 21/3/96:3, FG 22/3/96:1)

3 All existing controls in slaughter houses, meat plants and feed mills should be even more vigorously enforced.

(FG 22/3/96:1)

These regulations relate to the fact that the UK Government and its scientific advisors (the SEAC committee) located the source of the infection in contaminated meat and bone meal. Thus, it is suggested that offal meal became contaminated with scrapie, an ancient prion disease in sheep, when the rendering industry lowered the temperature at which such cattle feed was being sterilised. It seems that during the 1980's the UK produced 350,000 tons of meat and bone meal of which almost half was derived from cattle remains; sheep remains would make up about 15 per cent of the feed (ST 24/3/1996:1). It is the sheep content of the feed which was thought to have introduced the BSE agent. That, at least, is the official line on the matter.

Of course, there is also the small issue that feeding cattle remains to cattle not only converts herbivores into carnivores but also turns them into cannibals. According to the papers and television reports, the public certainly seem to think that there might be something deeply wrong with that particular practice. Moreover, the *Sunday Times* (24/3/96:13) informs us 'that warnings about the dangers of feeding animal proteins to herbivores . . . had been issued as long ago as 1979 when the Royal Commission on Environmental Pollution had concluded that such practices presented a risk that animal diseases might be transmitted to humans'. The same article also explains that the first government committee to report on BSE in 1989 which was chaired by Sir Richard Southwood highlighted the 'unnatural feeding practices' of modern intensive farming and concluded: 'We question the wisdom of methods which may expose susceptible species of animals to pathogens [diseases] and ask this general issue to be addressed' (quoted in ST 24/3/96:13). Not all scientific reports, it seems, are taken equally seriously by government, let alone heeded to or acted upon.

The extreme poverty of the bulk of press coverage and the narrow range of what constitutes 'acceptable scientific knowledge' becomes apparent when one takes note not just of reports on alternative interpretations of the source(s) of BSE and the extent of the crisis but, in addition, pays attention to 'Letters to the Editor'. In one such pertinent letter to the *Independent*, Professor Hugh Woodland from the Department of Biological Sciences at Warwick University points out that, despite the belief that there is a link between BSE in cattle and scrapie in sheep, SEAC curiously did not expect the pattern of transmission to be similar to that of sheep, that is, vertical from mother to offspring and horizontal between unrelated animals. He raises two further points about pastures and infectivity which seem to me to be crucial and concur with my own common-sense assessment of the situation.

In terms of a proposed slaughter policy, it is imperative to note that pasture that has carried infected sheep remains infective for some years after the sheep are removed. Slaughter alone [therefore] does not eliminate scrapie. . . . While the infected cattle are killed and burnt, this is only when they show recognisable symptoms. It is like the difference between Aids and HIV infection – the *asymptomatic* (my emphasis) phase provides the public health problem.

(IN 23/3/1996:18, Letters to the Editor)

As far as I could see (from my position as a citizen who consulted a wide range of papers and listened to a number of special programmes on the subject of the BSE–CJD link) neither of these arguments nor their implications have had the public discussion I thought they required.

In line with journalistic tradition, some dissenting voices and alternative interpretations to the official line have been presented in the press and on television. Thus, on Friday 22 March, 1996, for example, the *Independent* (22/3/96:13) published an article by Mark Purdy, an organic dairy farmer and BSE researcher who suggests an alternative source of infection to that proffered by SEAC and the UK Government. He points out that this same cattle feed was exported in vast quantities to farms across Europe and manufactured in the US without there being parallel outbreaks of BSE on the Continent or the USA. Moreover, 24,000 plus cattle born after the 1989 offal ban have gone down with the disease. What is unique to the UK, he argues, is the Government's requirement that dairy farmers treat their cattle for warble fly. The treatment involves a systemic pesticide that 'contains organophosphate – a chemical also found in a military nerve gas used in Iraq . . . and the basic unit of the infamous thalidomide' (IN 22/3/96:13). This chemical is designed to penetrate the skin of the animal so as to kill off the warble fly grub which could be found even inside the central nervous system. Mark Purdy would like to be assured that a thorough examination will take place of this potential source of the protein corruption in the brain of mammals. This was a powerful argument that stayed in many people's minds and it is astonishing how many members of the general public remember this particular story. Moreover, the hypothesis has resonated with many farmers' experience resulting in invitations for Mark Purdy to address farming organisations up and down the country.

A diligent reader in 'research mode' was able to find much argument, accusation, counter-argument and denial. Sadly, these debates were rarely to be found in the same newspaper, let alone the same edition. Thus, MAFF's counter argument to Mark Purdy's research was presented a few days later not in the *Independent* but the *Financial Times* in an article by Alison Maitland (27/3/96:9) who reports that Guernsey, which has the highest incidence of BSE in the UK, did not have the warble fly treatment for the island's herd (although, as I was to find out later, organo-phosphates are used as extensively on Guernsey as anywhere else). A spokesperson for the ministry was reported to have argued that it is the high proportion of sheep in the UK compared with other countries that is responsible for the BSE infection in the national herd: four sheep to every head of cattle, which meant a lot of potentially dangerous sheep remains in cattle feed. In the same article, Alison Maitland reports on a further counter argument when she recounts the UK Renderers Association's response to the Government's suggestion and the farmers' accusation that the lowering of the temperature at which the animal remains were rendered was responsible for the contamination of feed. The renderers argue in their defence that other countries instituted the same changes without having an equivalent outbreak of BSE. Like MAFF, they too argue that the large numbers of sheep in the country increased the potential for contamination. For Alison Maitland this raises the question of 'why the disease emerged in 1986, after the rendering changes had taken place, rather than at any other time'. This, however, becomes a strange question when we take explicit account of the time factor, that is, when we allow for the fact that BSE seems to take an average of four to five years' incubation before it develops into the full-blown symptom. For me, the argument about the high number of sheep leads instead to the question of whether or not the ministry and the SEAC committee have explored rendering practices and animal feed policies of other countries with a similarly high proportion of sheep among their farmed animals but no BSE epidemic in their cattle.

Careful horizontal study across the papers gave me some insights into the current state of knowledge on this subject matter as well as a very clear indication of the official line on the transmission of the disease across species barriers. In the first period after BSE day (from now on B-day) information was rather limited but, on the day after the announcement, the *Independent* (21/3/96:19) reported that in 1988 the Government's Southwood Committee (the predecessor advisory committee to SEAC) predicted that the disease would not be passed on to other species. Clearly, this prediction has not been borne out. For six years, according to the *Sunday Times* (24/3/96:13), evidence had been accumulating that BSE can and does jump species – its generic name after all is *Transmissible* Spongiform Encephalopathy (TSE). So far, we are told, the disease has been passed on to mice, cats, chickens, pigs, sheep, primates (marmosets), zoo animals and humans. In connection with this information, it was interesting to read front-page stories in the *Sunday Times* (24/3/96:1) and *Wales on Sunday* (24/3/96:1) about the worry of government advisors that sheep may have been fed with infected cattle meal and thus may now be re/infected. This, they conjecture, could lead to a BSE infection in sheep that was difficult to detect because it would be virtually impossible to distinguish the new BSE-based infection from scrapie which does not seem to affect humans. Thus, the fear is, first, that BSE in sheep would then move through the sheep population like scrapie – vertically between mothers and offspring and horizontally among unrelated animals

– which would make the disease exceedingly difficult to eradicate and, second, that in this form it would have the potential to affect humans in a way similar to BSE in cattle. For this reason, the *Sunday Times* informs us, a ban on ovine offal has never been closer.

A further important piece of the puzzle that helped to build up a picture about BSE was reported in the *Daily Express* (22/3/96:13). It stated that 98 per cent of cattle reported who have shown the BSE symptoms have been 5–6 year old dairy cattle. With that bit of information, of course, the thinking that young cattle should be slaughtered began to make sense. Again, however, the figures become confusing as soon as I related them to facts in other papers. An article in the *Financial Times* (23/3/96:1), for example, stated that, according to MAFF, 54 per cent of dairy herds and 15 per cent of beef suckler herds have had at least one reported case of BSE. The most important aspect of figures about cattle with BSE, however, is the fact that these numbers do not include the untold numbers of infected animals which currently incubate the disease but are being slaughtered before the tell-tale symptoms develop. So far, these animals are entering the food chain since no test has been developed to establish infection in live animals.

As a mother, I know that the majority of infectious diseases are at their most infectious before the symptoms appear, that is, at the sub-clinical stage. One wonders, therefore, why it is such a surprise to the UK Government and its scientific advisors that parents are transferring this knowledge to BSE and applying it as a precautionary measure for the safety of their children. It seems quite obvious to me that a test for live animals, administered to every animal to be slaughtered, is the only way to ensure that all British beef for sale in the shops is safe for human consumption. With that safety measure in place, the UK Government could apply the test systematically to the national herds, thus turning its attention to the eradication of the disease in cattle as well as sheep. If it was possible to develop such a test for HIV and AIDS, then why has no equivalent test been developed for the sub-clinical infection of BSE? Why, during the eleven-year period that this disease has been allowed to get out of hand, has there been no worldwide competition to come up with a test that establishes both BSE and its pre-symptomatic condition in live animals?

On Sunday 24 March 1996, four days after B-day, I took stock of what information I had been able to glean through the news mediation and realised how little I still knew. A huge tower of newspapers and many hundreds of articles have told me very little and what I had learnt was exceedingly difficult to extract. Moreover, I am convinced that a farmer, a beef exporter, or somebody from the slaughter trade would have come to the same conclusion, that is to say, the sort of information that was of importance to people whose economic livelihood was threatened was not forthcoming either, despite the fact that economic arguments dominated the reports. The hardest information to extract was anything relating to the historical background of BSE. Without the historical context, however, how can anyone make connections or understand the issues? In an effort to get some sense of historical context, I went back and extracted from across the breadth of the papers I had read during the first week after the announcement the following timeline of government action.

1979 – Warning by the Royal Commission on Environmental Pollution about the dangers of feeding animal protein to herbivores and its potentially harmful effect on humans;

1985 – First cases of BSE discovered and named in the UK;

12/87– UK Government officially recognises the link between feed made from infective bone meal and BSE;

7/88 – Government bans the feeding of cattle/sheep protein to cattle;

8/88 – Compulsory slaughter ordered for BSE infected cattle but only 50 per cent compensation offered to farmers (thus punishing honesty and encouraging lower numbers to be reported);

12/88– Milk from infected cattle destroyed;

4/89 – Southwood BSE research committee set up;

11/89– Bovine offal banned for human consumption;

2/90 – Full compensation paid to farmers;

7/94 – Ban of bovine offal from calves under six months of age;

3/96 – Ban on feeding animal protein derived from meat and bone meal to all farm animals;

3/96 – Carcasses from cattle aged over thirty months must be deboned in specially licensed plants supervised by the Meat Hygiene Service and the trimmings kept out of any food chain.

On the basis of this information, I am not surprised that papers supportive of the Government (which are the majority) were coy about recording in any explicit and comprehensive form the historical background since it demonstrates so blatantly the Government's minimal and inadequate action. One week after B-day, the *Daily Mirror* (27/3/96:4) published a time-line for its readers alongside a complete list of countries who had by then banned the import of British beef. Given the general confusion and lack of knowledge about the history of events, I was not surprised that this time-line contained somewhat different information and contradictory dates from the one I had been able to construct a few days earlier from the isolated morsels of information dispersed across the range of newspapers.

Reading horizontally across the papers has been instructive: not only did reports differ with respect to the interpretation of facts, but the so-called 'hard facts' did not match either. More importantly, it posed the question: who are the papers written for? In a national crisis such as this one where every citizen's health is potentially at risk, where the entire rural economy is threatened, and where the news media are the only sources of information, one would have expected newspapers and broadcasts to rise to the challenge of socially relevant analysis. They clearly did not. Press coverage over the months that followed, I am afraid, has not altered my opinion. From a media-adjusted level of expectancy it must be said that the reports in the *Financial Times* were a cut above the rest – but then, this is not much consolation to the readers of the tabloid press or the viewers who get their information from the main news channels on television. Investigative television journalism, in contrast, did take on board the challenge of this role. On Monday 25 March 1996 two programmes, *World in Action* and *Panorama* achieved what many hundreds of articles over the previous four days did not: they informed the public, posed pertinent questions and pointed the finger at a number of dubious practices as well as the Government's poor record on dealing with this ten-year old public threat.

In the remainder of this chapter, I want to reflect on some of the issues arising from what I have said above and on the way a health crisis has been re/framed by the media and Government as a beef crisis. I conclude with reflections on how the central role of social analyst is to be filled for future hazards of a comparable nature.

## Few reasons to be cheerful: BSE and mediated knowledge

As this book goes to press, BSE is still very much in the news. Over the preceding months, the British public and the people of continental Europe were drip-fed on a steady diet of disaster news. One-by-one previous scientific assurances collapsed: BSE jumps the species barrier; BSE is the most likely cause for the new strain of CJD in humans; transmission of BSE is confirmed to take place between mother and calf; the ban of 'dangerous' (that is, infectious) parts of the animal is extended to the whole head and the lymphatic system; there are first suggestions that the blood of affected animals too is infectious. I can only wonder where this leaves the UK's 'safe' muscle meat from dairy herds as well as milk – dairy products being one of the few home-grown products in which this country is self-sufficient. Day by day new scientific findings turn yesterday's safety into tomorrow's hazard.

### Sound science to the rescue

In this context of continuous scientific dismantling of previous truths, it seems incredible that the UK Government's belief in science's capacity to provide facts, truths and reliable predictions remains undaunted, undented and unscathed.

In 1988, the Southwood Committee which then advised the UK Government on BSE predicted that the disease would peak at 20,000 cattle developing the symptoms (IN 21/3/96:19) and have run its course by 1994. This was then 'best scientific advice'. With hindsight, of course, we know that the Southwood Committee's scientific prediction was way off the mark. Deaths of confirmed cases of BSE in UK cattle, the press inform us, are currently in the region of 160,000, which includes 24,000 calves that were born *after* the offal ban came into effect, and thus should not have been contracting the disease. Each time scientists present new findings, come up with new statistics and predictions, establish some novel connections and/or dismantle some old un/certainties, the current findings become gospel. In a bizarre farce of publicly displayed amnesia, each set of current data form the revised 'foundations' upon which the next round of statements of certainty and assurance are being built. Having dislodged the previous truths, each new devastating scientific finding in this tragic saga is swiftly reappropriated and celebrated as a positive step. Moreover, with a brazenness that is astounding, UK ministers demand that each new scientific result be taken seriously by governments the world over. EU decisions, insisted John Major, must be made on a *rational basis* and grounded in *sound scientific advice and evidence*. Scientific evidence, we are told, is the only sound basis for action. Each new evidence means now we know. Now we have certainty. Now we can act. That some of these 'findings' are statistical data based on models and a host of assumptions in the *absence* of 'facts' and certainty seems to make little difference to the deference for and inappropriate valorisation of science. In an article in the *Sunday Times*, Steve Connor and Michael Prescott explain this government-science relation in the following way:

> The scientists' conclusions had overturned a decade of government reassurances on BSE. The only way to handle the situation from thereon, therefore was for the Government to pin its colours to the scientists' mast. The policy was simple: we do what the scientists tell us, said one source.
>
> (ST 24/3/96:13)

This means, the UK Government has not only abdicated sovereignty to the global economy, as I demonstrated in the previous chapter, but at the national level it has handed over policy decisions about public safety to science. As always, however, the issue is even more complicated since, as I indicated at the beginning of this chapter, ministers have not been very meticulous in their heeding of scientists' advice. Instead, they have been highly selective and reformulated scientific recommendations to suit their political purposes. They have been turning statements about fundamental indeterminacy and uncertainty into assurances based on facts, talking about safety 'in the normal sense of the word' when there was and still is no basis from which to make such a statement, omitting warnings and recommendations wherever and whenever it seemed politically and/or economically prudent to do so.

The handling of BSE by the responsible ministers and their advisors can only be described as bizarre if one is reasonably kindly inclined towards them, scandalous if one has somewhat stronger feelings on the matter: they turned inherent uncertainties into factual statements and assurances whilst creating for citizens and businesses a context of utmost insecurity. Almost every week, the Prime Minister or one of his ministers promised a new date for something or other: a date by which the cull would begin, the cull would be completed, the ban would be lifted. Almost daily, the numbers and categories of cattle to be killed were being changed. Almost week by week, government officials moved the goalposts, leaving people who are economically affected by this disaster in a no-win situation. Since running a business is a temporally extended affair – something that has a history, requires planning, involves credit and interest payments – the Government's creation of uncertainty and the continuous changing of plans were the only foolproof way to inflict *maximum* damage on afflicted businesses. Ministers could not have done a better job if they had tried. Any person whose livelihood has been damaged or destroyed by this scandalous affair will tell you so.

Readers will see from these comments that I do not think there is much to be cheerful about. The sources for hope and the potential for action are few at the point of going to press, but there are some. These will become apparent as I reflect further on the issues: on the definition of the crisis, media knowledge without analysis, the mis-appropriation of time and, finally, on the potential for instituting social analysis for citizens, social theory for mass consumption.

### It's a beef crisis!

First, Stephen Dorrell informs the UK Parliament that the most likely explanation for ten young people having died from a new strain of CJD is that they had eaten meat that had been infected with the BSE agent many years earlier. The potential health hazard arising from this revelation could be of epic proportions, possibly unrivalled by any other in modern times. The BSE–CJD connection has the potential for developing into a gruesome modern-day equivalent of a cholera or pestilence epidemic with an unknown period of incubation. Second, the Government for whom the freedom of individual choice is sacrosanct informs its citizens that they had *no choice* over whether or not they wanted to infect themselves and their children with this horrendous disease. Red-meat eaters and even vegetarians are potentially afflicted. Understandably, this leads to widespread worry, concern, anger, uncertainty and a desperate need for information about the nature of this health hazard, its history, its current status, its future projection. Virtually none of this is forthcoming. Why? The public have misunderstood the statement, is the implicit government–media response. They have misinterpreted the message. They are suffering from a misconception. This is not a health crisis. *This is a beef crisis*. Those who think otherwise are wrong. The Government, the papers and the newscasters know: they have defined it as such.

Once the BSE–CJD link is framed in those terms and viewed from an economic perspective, a raft of implications follow:

1   Citizens' concerns about a potential health hazard are sidelined.
2   The Government can legitimately concentrate exclusively on the material task of rescuing a dire economic situation.
3   The entire issue can be dealt with from the firm basis of facts since (it is wrongly assumed) economics is not afflicted like science by the malaise of uncertainty.
4   People need not concern themselves with questions about the industrial methods of food production.
5   The UK Government, aided and abetted by the press – the tabloids more so than the broadsheets – can put their full efforts behind the effective externalisation of blame.

Anybody who has followed these events will recognise that this is precisely what happened. The health-hazard issue has been effectively silenced. It forms the velvet void at the centre of this tragedy, the nothing around which everything else revolves. *No action* is the UK Government's solution to this threat to public health. Let's wait and see how many people die of CJD in fifteen to thirty years' time: no test for BSE, no vaccine, no treatment. A potential hazard that does not arise as a symptom until decades hence quickly falls outside a beleaguered government's sphere of concern. This is/has to/ will be someone else's problem. There are far more pressing things to be getting on with in the here and now, not least to ensure a smooth passage to the next election. And, besides, none of the people taking non/action today will be around to take the blame in twenty years' time when the afflicted begin to show symptoms: far more effective, therefore, to take economic action which carries the promise at least of bringing more tangible results in a time-frame for which the current Government can still take the credit. Consequently, posturing in front of the European partners to get the ban eased and lifted, a few changes to the rendering industry and slaughtering practices, a promise on better enforcement of existing regulations, and the slaughter of unspecified numbers of cattle have been the key elements to the solutions of a redefined problem.

The press, obligingly, have gone along with this re/definition. They too have decided that it is a 'beef crisis': the banners that head newspaper pages and sections carrying reports on the BSE–CJD issue are labelled 'beef crisis', so are the identity tags attached to articles and their headlines. Almost from day one, this was the frame of reference through which matters around the BSE–CJD link have been analysed. It should therefore not surprise anyone that empty abattoirs as well as cattle and meat markets, lost export revenues, and restaurant and school menus dominated both press and television news headlines and coverage. The quite appropriate fear for jobs, livelihoods of businesses small and large, even entire industries for export and home markets have, accordingly, dominated the papers and constituted the bulk of articles written. Almost every day a new set of economic interdependencies was revealed, new clusters of relationships shown to be affected, further jobs reported to be lost. While many papers found it comforting to blame the EU and the Europeans for the UK's blight, papers such as the *Financial Times* showed how some other countries' beef industry and with it all the allied businesses and industries were in fact more severely affected by the crisis than the UK who was responsible for causing the collapse.

The press have colluded with this reformulation not because they were necessarily on the Government's side or were in some way tricked into this perspective. No, it so happens that such a reframing of the problem was a means by which to bring this rogue issue back within the fold of the familiar news world of reportable statements and events, describable disasters, and quantifiable economic facts and figures. It provided journalists with an effective way to report on the invisible, time-distantiated *Wirkwelt* of uncertain future threats in terms of tangible, material outcomes and the now–here world of political wrangling, that is, the politico-economic *Merkwelt* of the present. With the 'beef-crisis' frame of reference, the news media were back on familiar territory: answers to when, where, what, who, why and how questions could once more be factual, clear and unambiguous.

Furthermore, the 'beef-crisis' perspective enabled the news media to sidestep the challenge of informed analysis and avoid the need for continuous and unrelenting, careful and considered explanation of the nature and extent of the risk over a period of years, probably even decades. It meant journalists did not need to worry about historical contextualisation, wider social issues, arbitration between rival theories. The shift in perspective saved them from the challenge of engaging with something that is an ongoing process – indeterminate, immanent, in/visible, non/concrete, time-distantiated, latent, contingent, non/linear and not easily quantified – something that is timely without being recent and novel. Lastly, it shelved, for the time being, the need to address the issue of journalistic objectivity in a context where everyone is an implicated participant, where the outsider, spectator position has evaporated.

However, while this redefinition protected journalists from the inconvenience and excessively time-consuming need to rethink their role in the face of a national disaster, it also had its cost: it rendered ineffective the gallant efforts by many journalists (mainly of the broadsheets) to present alternative perspectives on farming and farming practices. I pondered long and hard about why these splendid articles did not work and lacked credibility. I am now convinced that it is impossible to tell a story of wholesome food and animal welfare from a 'beef-crisis' perspective that relates time with money, efficiency with speed, a country's economic health and wealth with GNP and export. (See also Chapters 2, 3 and 4.) Once the problem was defined in terms of beef sales and lost export revenues, the foundation had been taken away upon which a rethinking of industrial ways of producing food could be based.

## Time for reflection

### *The known unknowns and a few loose ends*

In the Viewpoint section of the *Farmers' Guardian* the following statement attracted my attention.

> The Government can be – and surely will be – blamed for all sorts of inaction. But it must be remembered that through it all the Government has insisted that it has acted on the best advice its scientists have been able to offer. It had no choice. To do otherwise would have made it party to some of the speculation that has fuelled the uninformed side of the debate.
>
> (FG 22/3/1996:1)

Is there anybody out there who has not engaged in speculation I wonder? In response to such heroic trust, it seems rather crucial that I briefly reconsider the basis from which this 'best advice' is proffered.

From a beef-crisis frame of reference it became difficult to enquire about the missing pieces in the jigsaw, because, of course, there were none. Best scientific advice had reassured the public that beef was safe and that this was all people needed to know. There was no need to worry about complicated matters outside their sphere of competence. The dangers were in hand. The only problem remaining for the UK Government and its mediators was how to save the meat industry and salvage the country's export prowess. However, if for just a moment of insolence, we were to suspend our belief in this factual world of the 'beef crisis' and allow questions of health and safety to surface, we would inevitably come up with a few unresolved questions and some inescapably 'unpatriotic' doubts.

From a health-hazard viewpoint, the key questions relate to public and personal safety, to hypotheses about the cause(s) of BSE and CJD, the transmission of BSE within and across species, the eradication of the disease, tests to establish sub-clinical infectivity, vaccines to prevent infection, methods of farming and food production. With this focus, of course, we are into the temporal realm of time–space distantiation, latency, and open future potential. From this perspective, indeterminacy and uncertainty are reflected back to us as soon as we ask even the simplest of questions: if BSE in UK cattle is due to contaminated feed then why do other countries who bought large quantities of this feed have no equivalent outbreak of BSE? If the infectious material is in the intestines (which we can assume on the basis that these have been included in the 1989 offal ban) then why are the pastures which are the recipients of much of their content not considered infected and infectious? If transmission is now 'known' to be between mother and calf, how is this taking place – in the womb, through the blood or through the milk?

Indeterminacy and uncertainty stare us in the face wherever we look and search for reliable information. No one, for example, can tell with certainty the incubation period of CJD. Accordingly, scientific estimates vary substantially. John Pattison, chairman of SEAC, suggests it might be in the region of five to fifteen years (IN 21/3/96:1). Stephen Dealler, a consultant medical microbiologist, thinks it is about fifteen years, which means the ten people who died from the new strain of CJD were infected sometime *before* the epidemic in cattle started (IN 21/3/96:2). Richard Lacey, professor of microbiology at Leeds University, the scientist who has most persistently and consistently warned about the grave dangers of BSE and whose gloomy predictions are one-by-one turning out to be mostly correct, suggests the widest range of incubation time. On the basis that scientists simply do not know, he proposes an incubation time for humans somewhere between ten and fifty years (IN 21/3/96:2). Equally unknown are the issues of dose and timing. (Dose and timing issues have been discussed with reference to hormone disrupting chemicals in Chapter 1.) In other words, scientists do not know whether the BSE agent acts like a traditional infection, a hormone-disrupting chemical, or in a further, yet undiscovered way. Scientists are even more in the dark when it comes to questions about multiple and/or interactive causes. Thus, ignorance reigns supreme, for example, around the question of to what extent dose, timing and cumulation are independent and/or interdependent variables. Furthermore, no reliable data seem to be available (as yet?) about whether or not the risk of infection is different for people of different ages and life stages, and whether or not it rises proportionally with periods of increased vulnerability.

Scientists may be able to tell us that the most likely cause of the new cases of CJD is infection through ingesting meat infected with BSE, but they are unable to tell us how and on what bases that infection has taken place. No one can tell with certainty how many infected cows – that is cows that are incubating the disease – have entered the food chain and the statistical models about this question are only as good as the assumptions and parameters that are fed into them. Even more uncertain is the potential number of infected humans: estimates range from a few hundred to half a million over a period of twenty years. Moreover, since no one knows at what point in the cycle of the disease an affected animal is at its most infectious, there is no way of telling which animals are safe to eat until such a time as a reliable test for live animals has been developed. Contrary to assurances, there is no certainty either about which parts of an affected animal are infectious. Instead, one 'safe' body part after another seems to be eliminated: brain, spinal cord and gut, the whole head, the lymphatic system, the circulatory system. Accordingly, there is a question mark over muscle meat and milk. They might be perfectly safe and then again they might not be. The fact that contemporary science lacks measuring instruments of sufficient sensitivity to establish infectivity in muscle meat and appropriate tests for infectivity in milk does not mean it does not exist there. Let me reiterate what I have argued in an earlier chapter. This situation has little to do with risks which are calculable: this is the post-Newtonian world of incalculable hazards. Journalists' questions about the size of the risk involved in this or that eating practice, therefore, miss the point; they misread, misrepresent and belittle the nature of the threat.

## *Institutional failure – what hope for social analysis?*

With the emergent health hazards associated with the BSE–CJD link, citizens have been confronted by tangible, multiple institutional failure. In the face of this inter/national crisis, neither the Government and its scientific advisors nor the news media have risen to the challenge of their public responsibilities. Instead, they have carried on their respective businesses as usual, which meant the translation of this intangible future-based health hazard into an economic issue and a matter of national pride, effectively silencing all and everything that threatens the *status quo*. While political hyperactivity and intense news coverage indicated the importance of the matter, non-represented non/activity and media silence do not mean all is well. They simply mean that other news have taken over, or that editors judged the story to have exceeded the public's interest span and thus to be threatening paper sales. The 'beef crisis', after all, still belonged to the long-term environmental issues that run on and on, and is therefore conveniently retrievable whenever the need arises: after ten years of existence, the beef 'scare' has an extensive history of oscillation between high profile attention and being kept out of sight.

> Suddenly at around 4 p.m. the phone stopped ringing. An eerie silence descended upon my office. BSE was no longer news, thank goodness. What was news? A cabinet minister, Mr Nicholas Ridley, had resigned following his offensive remarks about Germany. . . .
>
> The BSE story was relegated to the background. Members of the public who wanted to be reassured that beef was safe, were; they believed what they wanted to believe; and beef consumption might even have rallied a little.
>
> (Lacey, 1994:118)

Readily retrievable, the issue can always be updated from one now–here, a-historical and de-contextualised present to the next. Drip-feeding seems to be considered the appropriate method of communication for a public that is assumed to be afflicted by collective amnesia.

The need for change, as I suggest in this chapter, is pressing for the news media to fulfil their public duty as nationwide providers of information in accessible, that is, 'popular' form. For citizens to be served well in a situation of national crises such as the one that arose from BSE and the link to CJD, a number of parameters would have to be encompassed:

- The historical context would have to be covered extensively, cumulatively and in accessible form;
- The frames of reference would need to be de-naturalised. This would involve a discussion of the various frames of reference and their effects on what can and cannot be seen from the various perspectives;
- The silences would need to become the explicit focus of attention since it is these that provide the most illuminating insights into any topic. This would entail raising from the velvet void what tend to be the discomfiting features, illuminating what had been safely tucked away, out of sight, in soft and seductive darkness;
- Finally, it would involve engagement with the full multiplicity of temporal issues and their impact on the various institutional in/activities, with the in/determinate, contingent future of potential hazards, with the scientific uncertainty of the present, with the continuous and inescapable creation of outdated truths, the long-term, latent and immanent *Wirkwelt* and its relation to the tangible, factual *Merkwelt* of the phenomenal present. From a timescape perspective, it would be possible to discuss the disjuncture between the temporal horizons of a given hazard and the professional time frames of political and news media concerns.

With such a theoretical project, however, we would have left the journalistic world of the newsworker and entered the social theorist's sphere of competence. Sadly, social theorists are not renown for their down-to-earth communicative skills. The chance of reaching four million *Sun* readers are virtually non-existent. The question is therefore who and what needs to change? Which profession and which institution needs to make the dramatic adjustments required for communicative justice to be achieved for citizens? Or is it too late for adaptation? Do these new problems necessitate instead the creation of new institutions? In the face of the current overwhelming defence of the *status quo*, in the absence of any talk about something being amiss, and in the light of social theory's reputation of obscure elitism, I suggest that journalists and social theorists need to raise these issues together and confront the problems in collaboration, each taking steps towards rectifying their worst professional excesses.

As a first postscript note of optimism I would like to mention that, thankfully, readers do not necessarily believe all they read in the papers. There is, as Stuart Allan (1997) notes in a recent paper, no necessary link between the encoded message and its decoding. Despite or even because of the crass jingoism of the tabloid press, citizens are free to form their own opinions and judgements on the hazards, on the professional competence of those in charge of public safety, and on the validity of the media's interpretation and frame of reference. The second glimmer of hope is to be found in the change of public awareness: even the BSE–CJD fiasco has not been without its 'silver lining'. There are some, if very cautious, 'reasons to be cheerful' when we realise that arising from this public scandal is an increasing awareness that there are side-effects to the industrial way of life; that there are costs involved that need to be reconsidered; that treating animals like machines has consequences for the quality of their lives and ours; and that there are alternatives with the potential to become the new norm if enough people become sufficiently outraged and convinced that things need to change with regard to the hazards, the way they are handled, and the premise upon which social analyses are constructed and disseminated.

*Plate 6* Intensive cattle farming from Oct. '95 Focus by Jeremy Walker. *Source*: Tony Stone Images.

# 6

# RADIATED IDENTITIES

Invisibility, latency, symptoms – the case of
Chernobyl

## Introduction

In the age of nuclear power, members of industrial societies share the fate of radiation with a global community of beings for whom radiation works silently and invisibly from within. For them as for everybody else, radiation proceeds outside the reach of their senses: it is known only to their cells. Im/material and beyond the capacity of perception and sensibility, it affects our collective present and long-term future, our own and other species' daughters and sons of a thousand years hence. It is dispersed in time and space and marked by complex temporalities and time–space configurations. Its life cycles of decay span from nanoseconds to millennia. This means its time horizon too exceeds human capability and concern. Furthermore, radiation permeates all life forms to varying degrees and disregards conventional boundaries: skin, clothes and walls, cities and nations, the demarcation between the elements. Its 'materiality' thus falls outside the traditional definition of the real, outside a conventional view that has been absorbed as an unquestioned norm into the everyday understanding of 'Western' culture, the habits of mind where real is what is material and where this in turn is defined by its accessibility to the senses. Invisibility, vast variable time-spans of decay, networked interdependence and the fact that effects are not tied to the time and place of emission, therefore, make radiation a cultural phenomenon that poses problems for traditional ways of knowing and relating to the material world.

Nuclear radiation, as this book shows, is of course only one example of contemporary phenomena and processes that work invisibly beyond the capacity of our senses. Hormone-disrupting chemicals and genetically modified organisms are some of the other such industrially produced and induced processes that share these characteristics. All are recognisable only once they materialise through symptoms and once they have been identified through the mediating loop of science. Without the artificial sensory extension of laboratory science and medicine, therefore, people from all cultures and social strata are dependent on second-hand, thus collectivised, experience and knowledge for the identification and definition of such processes and phenomena. Thus, there arise questions about how to relate and respond to these in/visible yet central aspects of contemporary culture on the one hand and how to conceptualise and take account of them in the routine practices of everyday life on the other.

In this chapter, I focus on the nuclear accident at Chernobyl as an exemplar of invisible hazards that pose problems for some of the basic assumptions associated with the industrial way of life, the habits of mind that are rooted in Newtonian science, the linear perspective and neo/classical economics. To understand that event and its aftermath requires that we take note of the invisibility of radiation and the complex, contextual temporality that emerges from the disaster, its implications, and cultural responses. In order to bring to the fore the conceptual and practical difficulties that have to be faced in encounters with in/visibility and complex temporality, I first briefly restate some of the principle assumptions associated with the Enlightenment episteme that underpin everyday conceptualisations of reality in 'Western culture'. Next, I explore socio-cultural responses to Chernobyl and the associated im/materiality of radiation. In part three I consider the temporal complexity of the disaster and its long-term effects. This is followed by reflections on some implications for praxis of the mismatch between assumptions and the timescape of nuclear radiation in the context of the Chernobyl explosion and its aftermaths.

## Knowledge in the context of global/ising culture

The successes as well as the hazards of industrialisation, as I argue throughout, have their roots in a particular way of understanding and relating to the human and natural environment. In addition to the separation of culture from nature, the achievements and their associated dangers can be traced to the operationalisation of specific conceptual practices, most particularly the reduction of complex, interdependent processes to their component parts and functions, the isolation of things and linear event chains from their contextual interdependencies, and the imposition of the mathematical grid of a linear-perspective vision on the phenomena under investigation. This particular world view is strengthened by the conception of time as empty and neutral as well as its use as a standardised, quantified medium for exchange. What we today recognise as the unintended dangers of industrialisation, therefore, constitute the inescapable shadow of the successful operationalisation of that understanding. It is the darker down-side to such technological feats as mineral exploitation, the heat engine, electronic communication, chemistry in food production and medicine. As materialisations of a particular conception of reality, these products of scientific knowledge in turn feed back into the understanding of phenomena and processes. Deeply embedded in industrial culture, they shape habits of mind and thus determine and delimit imagination and theories, the hopes and fears that guide actions. Technology, argues Robert Romanyshyn (1989:10), is 'the enactment of the human imagination on the world'. Through it we not only 'create ourselves' but actually live 'our fantasies of service and control, our images of exploration and destruction, our dreams of hope and nightmares of despair'. As such, this technologically infused understanding is in need of being scrutinised for its historical relevance, whether it is still appropriate, that is, to the phenomena and processes that were built on its foundations but are now transcending its reach. A brief outline of these assumptions will suffice here as they have already featured in other chapters.

To begin with, the practice of reduction and the assumption of linearity affect our understanding in a way that makes it difficult, if not impossible, to grasp global/ising processes. To achieve the desired reduction, classical science relies on mathematical descriptions and technological metaphors. Their combined application banishes chaos, complexity, temporality, disorder, context, connectivity, and creativity from the analysis. Thus, classical mathematical description smoothes out irregularities, works with temporal symmetries where past and future are irrelevancies, and is understood to provide statements of truth and probability. In the mode of classical science, phenomena are explained through a scheme of cause and effect and represented by linear differential equations. Irrespective of the phenomena in question, the principle of proportionality applies: small changes produce small effects whilst big changes bring about big effects. Equipped with such conceptual principles, scientists (of whatever discipline) are not only able to generalise from one event and equation to another but also to predict future events.

Technological metaphors such as the clock, heat-engine, telegraph, camera, and the computer, in turn, represent an abstract world of interacting parts, a universe that can be taken apart and reassembled, a world of cogs, levers and reversible motion. Temporality – that is, change and the *Wirkwelt* – is eliminated from that vision through the theoretical development of reversible motion. Time is de-temporalised. As I have argued in Chapter 1, with the development of thermodynamic theory towards the end of the last century, time is re-temporalised but nature remains de-naturalised, lifeless: the temporality of thermodynamic theory refers to entropy and the unidirectional behaviour of dead matter, to its change towards heat-death, but it does not account for growth, creativity and the increase of self-referential order, key characteristics of living nature. The reductionist–linear vision thus represents nature in pieces and pieces of nature, a laboratory nature abstracted from context and the networked give-and-take of interactions.

Moreover, this traditional mode of knowing tends to eclipse the body. This is necessarily so because nature in general and the body in particular are intensely temporal, characterised by ageing and growth, creativity and self-renewal, by periods of intense activity interspersed with times of calm, by consciousness, feeling and emotionality – all those aspects of reality that have been negated by scientific reduction, denied reality status by linear causality and banished from the linear perspective vision. This mode of understanding, as I outlined in the first chapter, is effectively limited to the cultural products of human activity – the *Merkwelt* of its artefacts, technology, and institutions – and thus finds both the living, contextualised body and the *Wirkwelt* of *natura naturans* located outside its frame of reference and capability. As decontextualised spectacle and as specimen subject to fragmentation, mechanisation and quantification, body and nature are reinvented to suit the classical science mould. The Procrustean bed would be an appropriate image for this reconstruction of nature to fit the conceptual cage of classical science: reduction, simplification, causal relations, machine parts, and the window ethic, as discussed in Chapter 1. Alternative visions were offered by the Romantics of the previous century (again see Chapter 1); yet, the stresses between assumptions and subject matter were not yet of a level that would create sufficiently widespread public mistrust in that vision. Today, however, we find ourselves surrounded by events, processes and phenomena that place the classical system of thought under almost intolerable strain. As in/visible hazards, the symptomatic effects of the products of scientific technology have outgrown the limiting framework of their inception. In this and the concluding chapter, I use examples of the actual nuclear disaster of Chernobyl and the potential future hazards associated with the biotechnology revolution to illustrate not just the inappropriateness of these habits of mind but their hazard potential and dangers.

## Im/materiality and mediation of an invisible threat

On the morning of 28 April 1986 workers at a nuclear reactor 100 km north of Stockholm registered abnormally high levels of radiation, a reading which was soon to be confirmed by other stations across Scandinavia. These measurements were the first indication of an event that would send shock waves through the European nations and cast its long shadow over present and future communities across the globe. The Swedish measurements, though unmistakable and confirmed by others, could not provide (accurate) information about the source of the radiation; they merely indicated that 'something, somewhere, was releasing huge amounts of radioactivity in the atmosphere' (May, 1989:280). It was not until the next day that Moscow television issued a brief, three-line statement from the USSR Council of Ministers that informed viewers of an accident at the Chernobyl nuclear power station and assured the public that 'measures are being taken to eliminate the consequences of the accident' (May, 1989:280; McNair, 1988:140). Thus, whilst the Scandinavian measurements and others that followed were based on 'firsthand' information, everything citizens were to learn about the incident was based on retrospective constructions built up from a number of predominantly secondary sources: two Soviet (preliminary) reports to the International Atomic Energy Agency (IAEA) and the International Nuclear Safety Advisory Group (INSAG), a number of articles and television programmes by journalists, and analyses by social scientists and scientific specialists. (Much of this material is collated in a special two-volume information pack, edited for *The Ecologist* by Aubrey, Grunberg and Hildyard, 1991.) Similarly, knowledge of the consequences of the disaster is principally gained through the mediating loop of scientific interpretations and media representation. Thus, we learned that on 26 April 1986 reactor number four exploded in the Ukrainian nuclear power plant at Chernobyl. (For my brief account of the disaster I draw on Bunyard, 1988; Dörner, 1989:47–57; May, 1989:280–92; and the extracts in Aubrey *et al.*, 1991, which include

Aubrey 1991; *Energy Economist;* and Roche in *SCRAM*, 1990). The particular atmospheric conditions at the time sent a radioactive plume 1500 meters into the atmosphere (Bunyard, 1988:35) and diffused its radiation across the whole of Europe and beyond, falling as nuclear rain and contaminating all life forms at the level of cells in ways that elude sensory perception and scientific certainty. The outcome was, and still is more than ten years later, environmental destruction of enormous spatial and temporal proportions and untold numbers of *dead in degrees*: there and then, a bit later, tomorrow, and some time in some elastic future somewhere.

Interpretations about causes and reasons for the disaster varied enormously. Analyses differed with perspective and analytical framework, with the time-frame imposed on the event, with political and scientific interests, and with the degrees of implication and affliction (Medvedev, 1990). The explosion was, after all, the materialisation of the impossible. Retrospective reports on the Chernobyl accident were replete with comments about miscalculations, non-anticipated consequences, wrong predictions, over- and underestimations, profound and irresolvable disagreements between experts. Not surprisingly, the advice given to people in affected areas varied tremendously, ranging from maximum restrictions to silence which meant effectively 'carry on as normal'. Some affected populations were told not to consume locally produced milk and vegetables, some to stay inside, others to avoid drinking rain water. Some authorities withdrew from circulation all known affected produce whilst others simply raised the acceptable level of radiation (Aubrey *et al.*, 1991). These different and irreconcilable reactions demonstrate the difficulty of responding rationally and scientifically to complex, in/visible, im/material processes that are unbounded by future time and space. We can appreciate the nature of the problem when we consider with Ulrich Beck,

> What would have happened if the weather services had failed, if the mass media had remained silent, if the experts had not quarrelled with one another? No one would have noticed a thing. We look, we listen further, but the normality of our sensual perception deceives. In the face of this danger, our senses fail us. All of us – an entire culture – were blinded even when we saw. We experienced a world, unchanged for our senses, behind which a hidden contamination and danger occurred that was closed to our view – indeed, to our entire awareness.
>
> (Beck, 1995a:65)

In Kiev, Chernigov, and Gomel – all cities within a 100 km radius of the Chernobyl nuclear power station – the May Day celebrations were going ahead as normal on the day that the radioiodine content in the air was at its peak (Aubrey *et al.*, 1991:147, 150) when in other years the parade had been cancelled due to rain. The people of Pripyat, the town where most of the workers of the nuclear plant lived, were not evacuated until 26–36 hours after the disaster which means they were moved during the period of maximum height of radiation fallout. Another 90,000 people were evacuated after they had received between eight and eleven days of massive doses of radiation. Others as far away as 200+ km from the site of the accident were evacuated three and four years later because they too were approaching the new *raised* permissible maximum dose of lifetime radiation per person. Further afield, and more than nine years after the accident, some twenty farms in North Wales were still under quarantine and new 'hot spots' were still being discovered.

This difficulty of accepting something invisible as real applied to the unprotected workers involved in the reactor clean-up operation, to the mothers who nourished their children with contaminated food and sent them outside (to go to school, to play outside, to watch or take part in parades), to Government and Local Authority officials near and far away from the damaged reactor who failed to respond to the danger. It applied to the reindeer herders of Sweden, Finland and Norway and the sheep farmers of North Wales who had to be told by scientists and government officials that their animals and their land, their lives and livelihoods were contaminated by radiation. They all were and still are affected by something material they cannot see, touch, feel, taste or hear, something invisible that requires the extra-sensory perception of scientific instruments to accord it the status of material reality. Without scientific measuring instruments radiation is 'known' only to our cells and can be recognised only when it shows itself as symptom and thus becomes available as sense data. Its materiality is confirmed at the everyday level when people can *see* the dying children, the congenital deformities in all affected species, the mutated plants, orange-coloured pine forests without needles; when they become aware of the lack of bird song; when they experience the eerie sounds and feel the ghostly quality of deserted villages and dead forests. Finally, no one can know with certainty what is left in the entombed reactor and whether, in Zhores Medvedev's (1990) words, 'the radioactive volcano is really dead' (quoted in Aubrey *et al.*, 1991:150).

The in/visibility and mediation of this technologically constituted reality affect the sovereignty of the self, family, nation, species, earth.

> Without sovereignty over our senses, the dream of privacy is non-existent. Our notions of individuality, of self-determination of one's life, are founded upon personal access to reality. To the extent to which we are cut off from this access, we are driven – in the full flower of individualism – into a collective existence at the height of modernity.
>
> (Beck, 1995a:66–7)

Beyond the reach of the senses, radiation perforates the boundaries of person, species and earth and thus places humans and other life-forms in a new relation to each other, emphasises their communality. It extends to the depth of our being and the furthest reaches of our connectedness, from quantum self to the universe, from cell to brother pine and sister meson. As the implicate order of interconnections becomes recognisable, so spatio-temporal location loses definition without losing any of its importance. With the dispersal of radiation across time and space, in other words, there is no longer an unambiguously clear answer to the question, 'where and when am I?'. Moreover, there is no escape from involvement: observers, experts and analysts – the mediators of the Chernobyl disaster – were not merely describing objective facts from the disembodied neutral position of spectator outside the frame of reference. On the contrary, their presence and their reports *constituted* and *created* for a global audience an otherwise invisible and unknowable event. Through their presence and their interpretations, spectators were implicated in the 'materialisation' of that reality as it, in turn, settled invisibly in their cells. The window ethic described in Chapter 1 becomes an anomaly.

At one level, only what was written and communicated through the media became real, giving the event the textual quality and status that Derrida (1984) has written about with reference to nuclear war in 'No Apocalypse, Not Now'. At the local level, however, this 'textuality' was *experienced and lived* – in most cases with bitter and long-term consequences. In the immediate aftermaths of the explosion, two people died and many suffered severe burns and radiation sickness. Evacuated families lost homes and possessions and had to endure severe hardship before they were resettled 'in safety' just outside the 30 km danger zone. All are dying – in degrees – nothing textual about it for them. Those further afield, moreover, *live with the words* and their effects: agricultural produce destroyed; animals slaughtered for the fur trade or confined to specified, bounded parcels of land; livelihoods threatened. Depending on country and location within, the effects are interpreted differently and tied to widely varying policies and dictates. That is to say, the radiation, its source and its silent damage to the cells are the same, but the particular socio-environmental and political contexts with their associated temporalities – their latent *Wirkwelt* and explicit processes of change – play a crucial role in constituting the realities differently. Whilst scientists disagree over facts and prognoses and politicians play with 'safe levels', the people at the receiving end of the worst of the Chernobyl fallout 'know' that the cancers, the congenital deformities, the ill health and the premature deaths are the outcome of the explosion. Lack of scientific proof and the inability of experts to establish causal chains are irrelevant as far as they are concerned. Such incompatible perspectives and the mixing of traditional roles, however, are not unique to the Chernobyl disaster; rather, they characterise every nuclear and chemical incidence of this century, be they of a military, civil or mixed nature. Focusing on the contextual temporality of the event allows us to shift analyses from the Procrustean bed of classical science assumptions and dualistic conceptions – danger or safety, order or chaos, measurement or experience, dead or alive – to an understanding of complexity and implication.

## Temporal complexity as cultural practice

In the classical mode of analysis, events take place *in* time and time is utilised to establish their chronology – their date, duration and order. Moreover, in that form, time is employed unproblematically in the singular. It is used to measure change on a before-and-after basis and implicated in statements about cause and effect. Finally, questions tend to be raised about reversibility and cost associated with the duration of effects. This means time is constituted quantitatively and external to the event. It is operationalised by disembodied, 'no-body', experts and narrators. Time conceived in this way is ideally suited to complement the principle assumptions of classical science practice: of linear causality, abstraction, reduction and reversibility.

The nuclear disaster at Chernobyl and its effects, however, stretch that mode of analysis to its limits and show the inadequacy, paucity and poverty of that cohesive vision and its temporal assumptions: how could experts possibly identify causes when actions and reactions of operators and technology interacted according to different time scales, principles and goals, at different speeds and with incompatible timing; when symptoms were/are not tied to the time and place of the accident but permeate/d the globe to varying degrees; when futures are created then and now for thousands of generations hence; and when the periods of decay of radioactive materials stretch from a few seconds to millions of years? How could anybody establish 'effects' when those subjected to radiation fallout from the Chernobyl accident tend to show no visible signs of injury, when the power to destroy life and health is achieved silently from within, is transferred to subsequent generations and is quite unlike the wounds sustained in conventional accidents? Traditionally, cause and effect are linked in time and space – not so with the effects of radiation.

> Radioactive chemicals can now be found in the organs, tissues and bones of every individual in the Northern Hemisphere, and the contamination from past nuclear explosions will continue to cause environmental and health problems for hundreds of thousands of years, even if all nuclear activities stopped today.
>
> (Bertell, 1985:56)

How could scientists differentiate between radiation from the accident at Chernobyl, natural background radiation, radioactive pollution from nuclear testing, and leakages from nuclear power stations across the world? In a situation where the 'effect' is not confined to the region and time of its emission but pollutes water, land and air across the globe and where the entire earth community of living beings is implicated and affected to varying degrees for an open-ended period by the consequences of the industrial world's nuclear proliferation in general and the accident at Chernobyl in particular, the traditional scientific quest for proof and causes becomes inappropriate, and understanding of the relation between the past, present and future in need of revision. There arises the need to ask different questions, understand from a multitude of perspectives, and keep one eye on the future and the other on the past in a rather complicated way: to simultaneously peer into the future and the past whilst recognising the interpenetration and inseparable, mutual implication of what had been conventionally understood as the past, present, and future.

To be able to even begin to make sense of the cultural complexity of the Chernobyl disaster and its ramifications, time has to be liberated from its classical mould and understood in new ways. It has to be disentangled from its standardised, neutral, quantitative use and re-embedded and contextualised in the particular experiences, actions and meanings in question. From a timescape perspective, we appreciate that time, like the invisible radiation, is internal to processes and the *Wirkwelt*, and is thus not accessible to the senses. It too materialises through symptoms, that is, it becomes visible through the unidirectionality and irreversibility of ageing and growth, of explosion, fallout, and contamination, of mutations, deformations, thyroid complications, leukaemias and other cancer deaths. This in/visibility, im/materiality and internality, moreover, needs to be grasped in relation to the convention of using time as a standardised, external measure and exchange value. Not the replacement of classical time, therefore, but its transformation is the task (see Adam, 1990, 1995a). It needs to be conceived as a multivariate with each variant implicating all the others. At the very least, therefore, we would be concerned with time, temporality, tempo, timing, duration, sequence, and rhythmicity as the mutually implicating structural aspects of time.

Thus, with respect to this complexity, time at the most general level refers to the frames and boundaries within which we act and conduct our daily lives, construct chronologies and histories, establish biographies and mark key events. Time in this sense means the cultural convention of time measured in minutes, hours, days and years, of moons and seasons, of generations and periods of historical, archaeological and geological dimensions. It is also the external, standardised, historical and scientific framework under which we assemble and systematise some elements and eliminate others. As such, 'it is a means', as Bruno Latour (1993:75) suggests, 'of connecting entities and filing them away'. This conventional time of a standardised, external framework enables us to establish moment by moment the actions leading to the fatal accident, fix the time of explosion, chart the fallout as it moved across Europe, tie it to specific dates, and temporally map the pattern of its changing intensities.

A very different picture emerges, however, when we allow time to become coextensive with events and processes. Tied to position, time becomes temporal and the framework of observation relative. With contextualised time we can take account of differences in experience and handle variation in intensities and rhythmicity, in speed, pace, and pressure. Despite discernible overall patterns of contamination and damage, every person, animal and plant is affected differently. The existence of radiation 'hot spots', for example, means that there are differences not just between farms but even between single fields and areas within. Moreover, not just the symptoms but their temporal progression is unique: unidirectional, irreversible, but contextually unique. Moreover, time as function of position brings the seemingly objective spectators into the frame and acknowledges their implication in the subject matter. It is both a more honest and a more disconcerting conceptual/theoretical position to take: as participants in the midst of things we forgo the 'security' of contemplating events in tranquillity from some no-time, no-where, no-body position of historical and scientific objectivity. This shift in conception transforms the classical time as unitary framework and measure from sole contender to one framework among many, and it acknowledges the conventionality of its hegemonic position. The central importance of this shift, however, lies in the stress on agency and practice, in the realisation that every one of us and each of our actions count.

Temporality, as distinct from time, is established in the classical mode of analysis on the basis of before-and-after measurement. It is a particular conception of temporal difference where two points are established and the measured difference constitutes change: measure radiation on two consecutive days and the difference between the two measures marks the level or degree of change, thus its temporality. A second feature of the classical approach to temporality is the express belief in reversibility. Talk of 'undoing mistakes', 'reversing trends and decisions', 'getting the land back to what it was before' illustrates that particular assumption. A third characteristic is the understanding of processes in terms of cause-and-effect chains where input and output stand in a 1:1 proportional relation to each other: when you knock over a full wine glass you spill the wine and stain the surface, you might even break the glass, but the event is bounded in time and space.

The Chernobyl disaster calls these assumptions about time and temporality into question, necessitates their qualification, and requires conceptual revision. First, the before-and-after measurement could not tell us anything about differences within the time frame and it is silent with respect to anything that lies beyond those two points: it pronounces on the difference between two fixed points – nothing more. Even when lots of these measures are collected at very short intervals so that a semblance of process is achieved, the outcome is not temporality as an expression of the *Wirkwelt* but a string of fixed freeze-frames. Second, the experiences associated with Chernobyl make it both easier and more difficult to recognise reversibility as an illusion. On the one hand, the invisibility of radiation makes it hard to believe in the threat and its ongoing destructive action at the level of cells. As grass and plants begin to grow again the wishful belief in reversibility, that is, a return to a previously 'safe' condition, is all too tempting. On the other hand, however, the dying children and young adults in their prime, the deformities, the general deterioration of health and the realisation that this accident has cast a shadow over the future of people yet to be born shows the belief in reversibility to be a misplaced theoretical construct that bears no resemblance to the lives of people closest to Chernobyl who have been subjected to massive and intense doses of radiation. Further afield in the hills of Cumbria and North Wales the farmers too might have been guided by the hope that this invisible threat may be quickly washed into the soil to a depth where it no longer materialises on the surface as contamination, thus 'restoring' the land to its pre-radiation condition. But here too, the belief in reversibility has been shaken: radiation is proving to be 'stickier' than previously thought (Aubrey *et al.*, 1991:139–42). Farmers, consumers and parents have had to come to terms with the half-lives of radioactive materials – periods of decay that extend from the imperceptibly short to the unimaginably long – and with the material's variable passage through bodies. Not just the use but the quality of land has altered. Daily practices have had to be adjusted. Personal relations have changed: between affected farmers and those whose land escaped the critical dose of fallout, their surrounding communities, their local and national organisations. For many, their livelihood was (and in some cases still is) threatened to a point of no return. They 'know' that their land, their animals, their own bodies, and their relationships are *irreversibly* transformed: every aspect of their lives has been touched in a way that relegates the belief in reversibility to the never-never land of dreams and wishful thinking. Third, radiation is not bounded in time and space. It disperses systemically, permeates the material and living world invisibly, and materialises as symptom in un/predictable temporal and spatial positions. Fourth, the amount of emission is not proportional to radiation symptoms: in some species, individuals and areas of land, radiation works stronger and/or quicker than in others. Thus, when confronted with nuclear processes, the traditional way of understanding temporality loses its pertinence. It needs to be extended to embrace contextuality, irreversibility and non-proportionality.

An additional feature of the classical approach to time is its fundamental separation of time from temporality. That is to say, from the classical perspective you cannot focus on time and temporality simultaneously. Similar to Heisenberg's Uncertainty Principle, emphasis on the one means losing sight and definition of the other. To do justice to both aspects, we have to oscillate between them, deal with them in sequence – not so when we conceptualise time as a function of position and coextensive with events. In the latter case the complexity of times is implicated in that which is explicated at any one moment. Detaching time from temporality – the static forms from their forming – the classical mode of analysis associates time with culture and temporality with the processes of nature, which means it simultaneously separates culture from nature and the body. In contradistinction to these conceptual habits, radiation impresses on us the shared bases of existence on the one hand and the mutual implication of *Merk-* and *Wirkwelt* on the other. Threatening life at the level of cells, radiation ignores the boundaries between levels of being and species, nature and culture. It thus brings to the fore the networked connectivity of earthly existence, the unity of being and becoming, time and temporality, the link of the one to the whole, the common origin and destiny. With respect to time and temporality, therefore, nuclear processes unsettle the classical assumptions of dichotomy and proportionality, the dependence on cause-and-effect chains and the associated reliance on proof. Time lags, in/visibility, im/materiality, networked interconnectivity, and contamination in other times and places associated with the Chernobyl nuclear explosion make the classical theories of time and temporality unworkable.

Finally, the complexity of time is incomplete without the further incorporation of tempo, timing, rhythmicity and the interpenetration of past, present and future in any one moment of analysis. That is to say, when we focus on one of these aspects of time we must not lose sight of the others as important constituents of the subject matter under investigation. Thus, for example, any socio-scientific analysis of the Chernobyl disaster should not isolate the timing of operators' actions from other time factors, e.g. from decisions about the date and time of running the safety checks and tests which led to the disaster, with respect to the May Day celebrations and the predictable need for increased energy on the one hand and in relation to the fuel cycle on the other. Zhores Medvedev (1990, reported in Aubrey *et al.*, 1991:145), for example, points out that the test was conducted at the end of the reactor's first fuel cycle, which meant there were three billion curies of radionuclides in the reactor core which created additional hazards once things had begun to go wrong. Analysts need further to bear in mind the tempo, duration and sequencing of processes, actions and reactions.

For analyses of the responses to the disaster it is crucial to take account of the timings of specific newspaper reports, particularly those in the Western press and responding actions by Soviet government and local authority officials. McNair (1988), for example, devotes an entire chapter of his book *Images of the Enemy* to the timing of Soviet Press reports on the Chernobyl disaster. As I already noted above, the first indication of unusually high levels of radiation came from Scandinavian sources whilst the first Soviet statement did not appear until 72 hours after the explosion. The fact that the Western press was first to report on this matter placed the Soviet authorities in a defensive position, needing to respond to and counter an avalanche of ever more extreme Western media speculation. This situation of accusation, denial and counter accusation lasted until 6 May, ten days after the explosion, when President Gorbachov's *glasnost* approach seems to have won through and the first informative and self-critical reports began to appear in the Soviet media. The delay in providing as much information as possible in order to assist the authorities in their responses to the crisis, moreover, is centrally implicated in the way the disaster was handled not only in the Ukraine and in Byelorussia but also in the countries of Northern Europe that lay in the path of the nuclear cloud on the move. In Britain, for example, the more stringent safety measures were only introduced *after* radiation levels had fallen significantly, with the result that there was no longer any need to follow Germany and Holland in their ban of the sale of milk and other farm produce beyond, that is, restrictions on the slaughter and sale of lambs from affected areas for human consumption.

Thus, we can see, it matters when events happen: where and in what context, at what speed, in what sequence, over what duration, at what intensity, and under what conditions and pressures. Moreover, past and future are always implicated in decisions: past transgressions of safety measures, the history of animosity and mistrust between East and West, widely differing predictions and scenarios, recommendations about not having children, as well as the tragic future of generations yet to be born play an irreducible part in the Chernobyl disaster and its aftermaths. Chernobyl, like the other hazards I discuss in this book, therefore, demonstrates the need for new conceptual practices which not only include a timescape perspective but an altered appreciation of 'facts'.

## Relativity of measure, greying of 'facts'

The complexity, in/visibility and im/materiality of the Chernobyl disaster, its globalising tendency and perforation of spatial and temporal boundaries, and the inescapable need for mediation affect what is considered scientific fact. Collectively, these facets reduce the power of measurement, delimit the truth value of data, and severely curtail the utility of proof: black and white 'truths' shade into grey. Factual knowledge becomes partial, provisional and demarcated by its measurements, interpretations and representations. I mentioned earlier that only a small number of authorities acted fast enough to protect their citizens from the consumption of milk contaminated during the height of radioactive fallout. Because readings were not taken at the earliest possible point, milk was considered safe. The effects of this neglect have been felt across the world: contaminated milk powder from Britain, the EU and Poland found its way to Malaysia, Brazil, Nepal and Ghana (Aubrey *et al.*, 1991:141; May, 1989:290–1). However, it follows neither that these reported and detected incidences exhaust the cases of contaminated milk being sold and consumed, nor that they give an indication of the actual amount of contaminated milk in circulation: the measure gives no insight into the size of the problem. The same conclusion applies to the contamination of land and water, animals and people. Because not every square meter of land and cubic meter of water has been (or ever could be) measured, and because not every animal and person, present and future has been (or ever could be) examined for the full range of possible problems, and because measurements have not been (and could not be) conducted on a daily or even hourly basis before and since the accident, science has no means of establishing with certainty the extent of the problem. Citizens cannot even be sure that all the worst contaminated areas and beings have yet been identified. With complete cover of measurement an impossibility, scientists rely on mathematics, models and theory to establish 'truth' and 'proof' whilst expediency plays a substantial role in defining the boundaries of 'safety'.

The sensitivity of the measure, we have to appreciate further, depends on both its temporal and spatial scale. Traditional 'factual' and representational knowledge, however, is indifferent to scale. It was only through Einstein's Theories of Relativity that the importance of the time-frame of observation was recognised to make a difference at very high speeds. More recently, the inextricable involvement of scale began to be accepted for the observation for complex systems in the physical theories of chaos. (For discussions on scale and relativity, see Adam, 1995a:55–60, Hayles, 1990, Ingold, 1993.) These theoretical developments led to a subtle shift towards relativity and the realisation that the scale of measure influences not only what is visible but also what is considered to constitute a problem and what is deemed to be an appropriate response. That is to say, if we change the measure and classificatory principle we get different results for the same events. Thus, notions of 'facts', 'proof' and 'truth' clearly belong to the absolutist, representational framework of meaning used by theorists, politicians, lawyers and policy makers who continue to use the language of absolutes for phenomena that scientists have begun to recognise as fundamentally tied to the framework of observation and thus relative.

Acknowledgement of our dependence on the framework of observation, on scale and on measurement, however, makes the contamination no less real; it simply recognises its contested nature: John May (1989:284–7) gives a chilling account of the uncertainty and the lack of expert knowledge on the effects of Chernobyl; Crispin Aubrey et al. (1991) chart the conflicting expert opinions and document the widespread practice of doubling and trebling the levels of 'safety', whilst Ulrich Beck (1992a:23) points out that because risks and hazards depend for their existence on the knowledge we have about them, they can be 'changed, magnified, dramatised or minimised' and are thus 'particularly *open to social definition and construction*. Hence the mass media and the scientific and legal professions in charge of defining risks become key social and political positions'. The scientific becomes inherently cultural and political. Again, we have to recognise connections and interdependencies where classical theories work with abstractions and isolated phenomena.

In the light of these observations, what meaning can be attached to identity and the definition of safety? The perception of identity as unique and bounded by time and space is in need of revision when the damage of radiation, at the level of cells and bases of existence, is shared un/equally by all, when radiation disregards the boundaries of bodies, locality, nation and species, and when it both affects and is transferred to an unknown number of future generations. The effect on the meaning of safety is even stronger. In Chernobyl, in the days immediately after the accident, 'safety' began outside the 30 km exclusion zone around the damaged reactor: danger inside, 'safety' outside – until, progressively, safety levels of contamination were raised and safety zones extended. Four years later, the authorities were still identifying whole communities who had exceeded the 'safe' level of contamination and were thus in need of evacuation.

Even today the definitions of safety associated with the aftermaths of the Chernobyl explosion are still subject to revision. Yet, in spite of this, either–or assumptions still pervade official thinking whilst individual and society, nature and culture, safety and non-safety, contamination and non-contamination, knowledge and ignorance shade into each other, interpenetrate. Often experts and lay persons exchange their roles when local people give vital information to those in charge of defining and pronouncing on danger and safety. The black-and-white language of proof and certainty clearly has no place in the globalised world of nuclear power and radiation where death comes in degrees, a cell at a time, and where damage materialises in many guises – sometime, somewhere but in certain places more likely than others.

There is some predictability but this is based on principles other than those of classical science; it is not extrapolated from fixed before-and-after knowledge of the past but rooted in understanding of relationships and the inherent openness of processes. It is tied to developments that suggest nevertheless that some outcomes are more probable than others. From this altered perspective it is possible to recognise, for example, that risk analysis can inform about dangers but that the converse is not true. That is to say, low probability is not the same as making a statement about safety since anything not entered as data for the initial condition can pull the outcome in unpredictable directions. The assumption that risks are knowable, measurable and predictable – the basis of safety claims for nuclear power plants – clearly has been proven wrong and inappropriate with the Chernobyl explosion, which has shown radiation to be an incalculable hazard. Moreover, it has demonstrated that 'impossible' outcomes are not only possible but likely. The emerging 'perceived wisdom' recognises that nuclear reactors are *not* inherently safe but depend on a large number of active safety and support systems and it appreciates that our knowledge of hazards is in

principle incomplete because it is impossible to quantify, measure and predict the human factor, or to model and anticipate all possible event chains
and test all safety systems (Gruppe Ökologie, Hannover, extracted in
Aubrey, 1991:173). Despite the relative *measured* safety of nuclear power to
date, therefore, cultural practice in the nuclear age means coming to terms
with an inherent threat, now and for an indefinite future.

## Reflections

The explosion at Chernobyl has demonstrated the limit of classical assumptions. Because it was a social drama that involved, and still involves for an
open-ended period, real people making decisions, being heroic, suffering
slow deaths, grieving for their children, and fearing for those yet to be born,
the in/visibility and the im/materiality of the socio-environmental global
processes and their implications became real for all who know themselves to
have been touched by the disaster and all those willing to see. When the
control of atomic energy went so catastrophically wrong, the taken-for-
granted habits of mind became visible and the conceptual tools of classical
science came tumbling down with the explosion: the insistence on proof
based on unbroken causal chains, the expectation of proportionality, the
dependence on sense data and thus materiality and visibility, the reliance on
objectivity and neutrality, the habit of abstraction and reduction, the search
for simplicity, the status of laboratory knowledge, and the separation of
culture from nature. Networked, global connectedness is demonstrated when
action by workers in the Ukraine can threaten the livelihood of farmers in
North Wales and when, in turn, their milking of cows radiates babies
in Malaysia. Moreover, Chernobyl pressed home the shared basis of existence:
in the threat, the communality of vulnerability and needs was inescapable. It
made visible the unbroken wholeness across space and time, the im/material
*Wirkwelt* processes below the surface. It showed us not as observers but
implicated participants whose actions matter because they construct reality –
here and now, everywhere and for all time. It demonstrated that nuclear
power is an international issue. A hazard that knows no boundaries has no
place in nation state politics. Its status of national sovereignty and security is
misplaced (see also Chapter 3). Global cooperation not competition seems to
be the only appropriate interactive principle. *Cooperate or perish* should
therefore be the slogan for the radiated 'nations' of the nuclear age.

*Plate* 7 Sheep, Co. Kerry by Giles Norman. *Source*: The Giles Norman Gallery.

# 7

# GENIES ON THE LOOSE: WHAT NOW ALADDIN?

## Introduction

With genotechnology all the issues I have raised in this book come to a head: for science, indeterminacy reaches a climax and the hazard potential is mind boggling. The involvement of the world's major transnational companies is virtually total. The difficulty of national governments to arrive at policy and regulation is compounded by the global sourcing, the complexity of the science, and the networks of interests that span most spheres of social life. For farmers the control over the mode and means of re/production is moved even further out of reach: intensified is the squeeze applied by the agri- and bio-chemical companies that supply them, a science that defines for them what is modern and up-to-date practice, and ever larger debts to service the system. With respect to getting the all-important media coverage, genotechnology is devoid of any features that would make it attractive for popular media attention. And, finally, for citizens the potential hazards and social problems are amplified: the poorest people of the world and the members of countries who cannot afford to apply strict regulations are the unsolicited guinea pigs of this massive experiment. To the extent that this is largely a reproduction and food issue, women the world over find themselves extensively implicated. Third World producers are deeply affected by the corporate global pursuit and utilisation of this technology whilst people with genetic diseases are at the sharp end of its application in medicine. Consumers are expected to eat the food in blissful ignorance. The recipients of genetic screening are to be thankful to be told their potential genetic fate. In the absence so far of any spectacular disasters, complacency reigns supreme.

In this concluding chapter I use the focus on genotechnology to revisit some of the areas of concern that emerged during the course of this investigation into the timescapes of contemporary environmental hazards. Instead of bringing the book to a conclusion, however, I open up the issues further and deliberately leave them unresolved in the hope that this will instil a sense of urgency and discomfort befitting the contemporary condition. I look at the science of genotechnology, its assumptions and hazard potential against a background of promise and hope, offer some thoughts on the interests involved, and reflect on this particular socio-scientific creation of the future. For my understanding of this subject matter I have drawn primarily on the following books and journals: Sheldon Krimsky (1991), Sabine Rosenbladt (1988), Vandana Shiva and Ingunn Moser (eds) (1995), and *Politische Ökologie* 35 (1994) on the theme of genetically engineered food; as well as the writings of Regine Kollek (1995a, 1995b) and Peter Wheale and Ruth McNally (1988, 1995).

## The eighth day of creation: technology with green credentials?

Modern biotechnologies have truly revolutionary implications. They confer a *generic* technical capacity to undertake selective genetic intervention in existing forms of life and to create novel life forms. The genetic code can now be manipulated and nature refashioned according to the logic of the market place. As a result, biotechnologies have suddenly new alternatives and paths of development for all major actors in the food system: farmers and input suppliers, primary processors, final food manufacturer and customers. It is this crosscutting, polyvalent capacity of biotechnologies which makes them such a potent force in the food system.

(Goodman and Redclift, 1991:167–8)

This kind of optimism about the potential and promise of genotechnology is repeated in the field of medicine where a parallel and similarly wide-ranging development is taking place. The genome project which seeks to map the entirety of the human genome, genetic diagnosis of hereditary diseases, and the potential for future gene therapy excite not just medical scientists but most of the major transnational bio-chemical companies, the financial markets, and the world of insurance. Numerous collaborative ventures between academic scientists and business have evolved, making some of the participants millionaires before they have brought a single product on the market. Moreover, all the major chemical giants are involved. Thus, for example, Zeneca, the bioscience arm of ICI, employs 33,000 people worldwide and, according to Geoff Tansey and Tony Worsley (1995:176), in 1992 had sales of nearly £4 billion. The scientific interaction with the basis of heredity is big business: the stakes are high, regulation meagre, public information virtually non-existent.

Genotechnology is the latest in a string of scientific developments that extend industrial societies' uncontrolled reach into the very, very long-term future. Like the chemical and nuclear technologies before it, genetic engineering is bringing about substantial changes in human existence which extend well beyond the time–space confines of industrial societies. It affects the future of glocal public health, agriculture and food production, trade and export, insurance and financial markets and, as patenting and the seed monopolies take root, the livelihood of peoples the world over. It implicates city dwellers and contributors to the rural economy, members of industrial and tribal societies. It has an impact on peoples of both the Northern and Southern hemisphere. Like previous technological innovations, it holds out the promise of cornucopia: the end to food shortages and world hunger, poverty and disease, weather and season dependence, as well as the beginning of limitless supplies of food, new forms of speciation and genetic diversity, control over the processes of life. As with the preceding technologies, the promoters of genotechnology focus on the promise of profit and control at the expense of issues associated with the potential long-term effects, on function and product at the expense of re/productive system processes – the visible *Merkwelt* without the invisible *Wirkwelt*. As before, therefore, the tip of the iceberg is mistakenly assumed to be the whole.

### Genotechnology to the rescue: the promise of cornucopia

In the context of the discussions presented in this book, this newest and most lucrative branch of scientific research and its products impacts on glocal society, inter/national politics, economic activity, medicine, and agricultural practice in a number of significant ways: first, this technology relates to social stratification and the creation of differential life chances. Genotechnology, it needs to be appreciated, is redrawing the lines of inequality. 'Firms involved in biotechnology', as Geoff Tansey and Tony Worsley (1995:177) argue, 'promote a new form of enclosure, no longer of common land but of the common heritage of humankind in the genes of plants and animals which are needed in much genetic engineering work'. By patenting the methods and procedures of genotechnology, the owners of patents have effectively negated the farmers' ancient right to the means of reproduction. Saving seeds from one year to the next and thus ensuring the best growing potential for specific localities, for example, is no longer an option: the introduction of hybrid seeds meant that growers cannot do it; the patenting of genetically engineered seeds means that they are not allowed to.

This shift in practice marks a move from sustainable self-sufficiency to one of dependency on the products and services of agri-chemical and bio-tech companies. The transcendence of self-sufficiency, a second effect of genotechnology, is in fact one of the proud achievements of this technology: the mode and means of reproduction have been effectively appropriated by science and business.

> Even though it has been a tradition in most countries that a farmer can save seed from his [sic] own crop, it is under the changing circumstances not equitable that a farmer can use this seed and grow a commercial crop out of it without payment of royalty . . . the seed industry will have to fight hard for a better kind of protection.
> (Hans Leenders, [General Secretary] World Seed Houses, quoted in Shiva, 1993:122)

> No company would be willing to part with what they took years and spent millions of dollars developing. It's a question of intellectual property rights.
> (Jan Nefkins, Cargill South-east Asia Ltd., quoted in Shiva, 1993: 125)

Each and every generation of plants and livestock together with appropriate chemical aids have to be purchased from their proud developers and owners – from Monsanto and Ely Lilly, Zeneca and Shell, Upjohn and Bayer, Hoechst and Ciba Geigy, Nestlé and DuPont – not to be reproduced but used, consumed and/or discarded (see also Chapter 5).

Genetic engineering is therefore an issue of time-based control not so much over the mode and means of production but even more so over the mode and means of reproduction. The patenting of seeds, semen, plants and life processes, fertility treatment for women, and genetic testing in medicine all point in the same direction: the attempt to control the future through reproduction. Loss of self-sufficiency and power over the beings, people and processes involved happens to be one of the key by-products of this control.

A third impact of genetic design in medicine, agriculture and food production relates to the econo-political ideal of choice. Much effort is currently expended to force producers to declare all genetically engineered materials fully, especially those that enter the food chain. The issue of choice, however, is not exhausted by the indisputable need to have genetically modified products clearly labelled. Money and knowledge determine whether or not citizens are able to exercise choice. In Chapter 5 I have made the case with reference to money and the capacity to choose healthy foods. Here I want to add the issue of knowledge. Since choice presupposes appropriate knowledge upon which to make decisions between alternatives, proper choice about labelled genetically engineered foods, genetic screening, and future genetic therapy can only be exercised after a substantial level of gene literacy has been achieved in any given population. In a context of widespread ignorance, in other words, choice is rendered meaningless. Loss of choice is thus inescapably tied to the genotechnology rhetoric of choice. There is a need, therefore, for people to become gene literate and for this knowledge to be as widely available and socially dispersed as computer literacy is today.

A fourth effect relates to the issue of scientific control. Genotechnology as the crowning glory of biotechnology developments brings to fruition the quest for technological rationalisation and control of all spheres of socio-environmental life. That is to say, with genotechnology science has extended its reach beyond the breadth of social life and natural phenomena to the depth of the life force of nature. It has transcended the product barrier and now operates at the level of nature's processes. This means scientific exploration is no longer tied to the *Merkwelt* of visible outcomes – the sphere of *natura naturata* – but has extended its investigative prowess to the *Wirkwelt* of nature's immanent processes – the realm of *natura naturans*. That is to say, with gene manipulation, scientists are able to interact directly with the source of life and change. As such, genotechnology could be a means of closing the gap between technology and living processes. This, however, is not the case. Far from closing the gap, this paragon of Newtonian science, as I shall indicate below, turns the crevice into an unbridgeable chasm. The time-based critique offered throughout this book, therefore, becomes more pertinent than ever.

The image and rhetoric associated with this technology is clearly very different from the four effects I have just mentioned. Not the intensification of inequality, profit potential and control – over nature's processes, agricultural means of reproduction, and citizens' capacity to exercise choice – but the horn of plenty, increased diversity and biocentricity are the publicly hailed benefits of genetically engineered products. Cornucopia is the promise, especially where genotechnology is applied to the production of food, seeds and animal feeds. Science is to provide the technological means by which the potential is to be materialised, by which humanity's needs are to be furnished. That is to say, science in the form of biotechnology and genotechnology tends to be promoted as the panacea to most of the major problems of twentieth century human existence: to hunger and food shortages, pollution and environmental degradation, the reduction in biodiversity, disease and even ageing. This technology, moreover, is presented as the scientific knowledge revolution that can repair the damage wrought by industrial activity associated with the use of machines, fossil fuels and synthetic chemicals. It has already produced micro-organisms that devour oil slicks; plants that thrive in communion with pesticides, herbicides and fungicides; foods that taste fresh after months of transportation and storage.

This technology, moreover, so the official line goes, is not merely the mode and means to solve societies' problems but it can deal with them in a safe and nature-friendly way. It is the soft, energy conscious path to a sustainable future. In the words of Earl H. Harbison, Jr., president of Monsanto which is one of America's foremost transnational companies involved in the development of biotechnology, 'we can at last find biological solutions to biological problems that mechanization and chemicals cannot solve' (quoted in Krimsky, 1991:84). Biocentrism, sustainable growth and a technology in harmony with nature are, however, difficult to achieve within the competitive system of global trade, as discussed in previous chapters. Moreover, it does not square with the view of nature that arises from the scientific knowledge associated with this technology.

### Biocentrism in a context of de-naturalisation and economic efficiency

Where transnational biotechnology and genotechnology companies wax lyrical about their green credentials, their biocentrism, and their environmentally sensitive way of exploiting nature, the scientists at the forefront of this development dream of control, mastery and the creation of nature to human design. Biocentrism as the *modus operandi* that is most in tune with nature's own ways is quite definitely not on molecular biologists' agenda. Nothing could be further from their intentions. On the contrary, a biocentric approach would bring their work to an immediate and abrupt halt: genetic engineers would have to reconsider their positions. That is to say, the science that creates the basis for commercial exploitation of the deep-structure processes of nature is by definition intensely and exclusively anthropocentric: its aims are, after all, the *improvement* and the *redesign* of nature to human desire, requirement and convenience. Its vision is not one of harmony with nature but of an 'eighth day of creation', of taking over where god(s) and evolution have left off, of gaining knowledge and control, of improving on existing designs, re/creating nature for the exclusive needs of humans and their descendents. This human-centredness applies whether the objects of research are animals that produce human proteins for pharmaceutical companies, genetically altered pig hearts for transplantation into human beings, pigs with an extra rib to provide more chops per carcass, a growth hormone that increases the milk yield per cow per temporal unit of measurement, a food crop that can be produced outside its natural habitat, or a recombinant virus that is to act as a vaccine against rabies in foxes. Every single development is exclusively for the purpose of human convenience, well-being, and/or profit enhancement. Questions about species being or the needs, rights and well-being of non-human life forms as well as nature and the environment more generally are definitely not under consideration here. They do not feature on the agenda of either molecular science or the biotechnology industry.

On the contrary, cornucopia, the future state of plenty, is achievable only by an explicit distancing from what is 'natural'. Through intervention at the molecular level, genotechnology accomplishes precisely that which is *not* possible in nature: the recombination of genes across breeds and species; the development within species and breeds of characteristics and capacities that do not and are not likely to occur naturally; the production of food outside its 'natural' context of existence, that is, independent from soil, water, photosynthesis, sexual reproduction, and the rhythmicity of seasons (see also Chapter 5); the test-tube creation of body 'parts' which means their production outside living cells and bodies; the 'cure' of diseases before they happen. Sheldon Krimsky describes this denaturalising process with reference to the production of food:

> Food will be viewed as the output in a production process that changes a feedstock into proteins, vitamins, and carbohydrates, and creates a product that people or animals either require for growth and nourishment or desire or would consume. There are many ways that these conditions can be met without farming the land. . . .
> The fermenter, bacteria, engineered genes and feedstock would replace soil, sun, rainfall, and germ plasm as the primary factors in production.
>
> (Krimsky, 1991:10)

Not harmony with nature but distance, decontextualisation, disembodiment and the transcendence of nature's evolved processes of speciation and reproduction are the goal of this technology. Thus, instead of a science in tune with nature, for Jeremy Rifkin (1985: 74), 'genetic engineering represents the ultimate negation of nature'. Only in this form is it of interest to industry. Only from their 'unnatural' bases can genetically modified products develop into ultimate expressions of industrialisation.

Furthermore, not nature but the factory and industry are the metaphors and models that guide research associated with genetic manipulation. There is talk of 'gene machines' and genes as 'survival machines', of cells as 'chemical factories' and proteins as 'building blocks', of speed-up, efficiency and productivity. The machine metaphor indicates interchangeable parts – genotechnology rearranges bits of genetic information. *The cell as the material basis of life becomes the source for a new bacterial mode of industrial production.*

> Industrial scientists study how to amplify or enhance the cell's products by introducing multiple copies of a gene by controlling environmental factors. As a 'factory', the cell must be brought into peak efficiency.
>
> (Krimsky, 1991: 6)

In food production, such 'efficiency' is linked to an optimal conversion ratio of input to output such as grass and/or feed stuff to meat. In factory farming, this ratio has now reached its peak and is currently in a phase of diminishing returns where marginal increases in efficiency have reached a plateau. In the history of agriculture, each time such a phase had been reached, new technologies were invented to inject a boost of profitability: mechanisation, chemical assistance, economy of scale, and now genetic engineering. In this latest phase of intensive agriculture, the morphological features of animals and plants are redesigned so as to yield a more profitable conversion ratio. Examples of such genetically engineered efficiency would be fowl consisting primarily of breast meat and crop plants with accelerated, extended or additional growth cycles for the more efficient use of land. In order to fully appreciate the significance of this technology, we need to grasp some of its defining features and the specific temporalities they entail.

## Of mastery, omnipotence and the hazard potential of success

In everyday understanding, genes are associated with the part of heredity that is pregiven and predates individual existence; they thus stand for that which cannot be altered by upbringing. Molecular biology since the 1970s, however, has fundamentally altered, even negated this preconception. Far from being fixed, genes have become the sites for change of what was previously considered impossible even within the time frame of evolution: with the recombination of DNA sequences genotechnology has been able to transcend breed and species boundaries. Context has been rendered irrelevant. Quantity is no longer tied to sexual reproduction. Several changes can be achieved simultaneously rather than sequentially.

Genetic engineering as the deliberate manipulation of hereditary material is, of course, not a new but an ancient activity. As selective breeding, it has been practised almost as long as humans have been domesticating animals. What differentiates contemporary molecular genetics from this established tradition is its capacity to effect change at the level not of the phenotype but of individual genes. With the development of this radically new set of techniques scientists are able to intervene at the very source of being by decoding, comparing, excising, splicing, recombining, transferring, and cloning individual genes and sections of DNA. This scientific manipulation at the level of genes opens up the foundation of life to human intervention and design. It allows for the injection of characteristics that are not part of the hereditary genetic make-up and enables scientists to combine morphological and functional characteristics that have evolved separately for millions of years and therefore do not occur anywhere in nature. In their synthesis these human creations constitute truly new forms that can be replicated at will and in unlimited quantities. At that level, therefore, scientific dreams about omnipotence have become reality. At another level, however, the loss of control is unprecedented. The novelty of genetically engineered life-forms and their transcendence of nature's evolved processes and complex temporal interdependencies are also the basis for the magnitude of their inherent hazard potential. Thus, the increase in scientific control seems to stand in an inverse relation to the loss of control over effects.

The work of Regine Kollek (1995a, 1995b), a German molecular biologist, is helpful for understanding this complex interdependency. The scientific manipulation of genetic material at the level of the genotype, she explains, is only possible because way back in the evolution of life-forms all organisms share genetic origins and are thus genetically related. This means that even very distant species share nucleic acid sequences or functional genes so that specific proteins and enzymes, for example, may be found in the cells of yeasts, insects and mammals. Only on the basis of this shared evolutionary prehistory, therefore, is the transfer of genes from one breed and species to another possible. What is crucial to note, however, is that proteins or enzymes with identical biochemical properties do not necessarily fulfil the same function, that is to say, the same material property has, depending on its specific context, a different ecological relevance. The invisible time–space of interaction is central to particular material expressions. The function of a specific gene, therefore, is not only given by its DNA sequence but by its 'location within a particular chromosomal and cellular context' (Kollek, 1995a:100–1). Moreover, beyond this internal influence of cellular context and position, phenotypes are developing with reference to their extra-cellular contexts in interaction with their environments. This means that any genotechnological intervention has ramifications at all these levels.

> Parameters related to space, time, biology and natural history which influence the characteristics of individual organisms as well as the way they interact with other organisms, and which have proven themselves to be useful, perhaps even necessary, to life on earth, are therefore altered by genetic engineering techniques.
>
> (Kollek, 1995a:98)

The controlled conditions of this experimental science create a context within which to produce specific, desired and highly predictable results about the action and function of *individual genes* only. Since, however, the multiple layers and levels of interactions and contextual differences do not form part of the experimental paradigm, the embodied, embedded phenotypic and environmental outcome is anything but predictable or certain. There exists no comprehensive theory to date, argues Regine Kollek (1995a: 101), that is 'capable of describing the relationship between the functional effect of a gene and its spatial arrangement within the genome'. This means that the timescape of genes is only indirectly encoded in DNA and cannot therefore be deduced from the structure of individual genes.

Moreover, since interaction, interrelationship and socio-temporal context do not form part of the theoretical framework of the experimental science of molecular genetic engineering, its knowledge is only valid for the laboratory context. It does not apply for the interactive reality of phenotypic involvement in the environment. In other words, the behaviour of a genetically altered organism in its environment cannot be deduced or predicted from its controlled laboratory conditions. Thus, whilst releases of genetically modified organisms, for example, may not necessarily result in a hazardous situation, the efficiency of the experimental manipulation nevertheless inescapably stands in an inverse relation to its predictability in the environment. 'Uncertainty and risk', Regine Kollek (1995a:103) therefore suggests, are 'the price which must be paid for the total accessibility and control of these living objects in the laboratory'.

This genetic work at the 'leading edge' of science and technology with its associated creation of financial profits is thus inextricably tied to indeterminacy and, as I show below, the potential for hazard creation. Dreams of certainty and control are simultaneously within reach and utterly impossible. Mastery and omnipotence, in contrast, are unlikely for even the most tightly delimited laboratory context. The everyday understanding which associates genes with the pregiven may not, after all, be fully negated. Focus on the temporal complexity of genotechnology gives us an indication why this might be so, why certainties and assurances need to be re-scrutinised, and why both theory and practice of the industrial way of life are in need of revision.

The timescape perspective enables us to see the invisible. It facilitates recognition that the hazard potential of genetic engineering is inescapably and proportionally related to this technology's successes. The accomplishments, we need to appreciate, are staggering. But, then, so are the perils: millions of years of co-evolution are circumvented. Reproduction cycles are dramatically speeded up or cut out altogether. Here the extraordinary achievement is counterbalanced by the inability to observe when things go wrong and to stop at each of the stages along the way. The achievement thus entails the loss of the established method of trial and error. Although the unprecedented speed-up of processes has obvious economic advantages, it clearly contradicts some very important scientific principles. Instead of exercising caution and precaution, genetic engineering is, in the imagery of Colborn *et al.* (1996:246), like piloting a jet plane with a blindfold on.

Finely tuned, interdependent processes that can be abstracted from their context and set aside to be speeded up and slowed down at will allow for control over time and the temporal processes of nature. In a social system where time is money, such control is a tremendous feat. Its pursuit is irresistible, the potential for huge profits persuasive. Since, however, the ongoing system processes of interactive, mutual effectivity are unbounded, open-ended, long-term, time-distantiated, and often with long periods of immanence and latency, the hazard potential is matched only by the scale of the potential financial gains.

With the transcendence of the rhythmic and cyclical reproductive processes of life and death, growth and decay, production is freed to a substantial extent from its dependence on seasons and climate. Yet this achievement too is realised at a cost. It entails the transformation from an ecological, no-waste, recycling, reproduction system into an industrial linear, one-use, input–output production structure in which neither the production residues nor the products after their use are reabsorbed as productive forces into the system. They become 'waste'. As waste, the material not only contributes to the overall entropy of the system but it also interacts with its environment and moves invisibly through the networked system: unplanned, unsupervised, and uncontrolled. This current genetic control over life and death processes therefore is inextricably tied to the dramatic rise in hazard potential now and for the very long-term future, to the shaping of indeterminate, hazardous futures. This intensification of hazard potential needs to be acknowledged as real even though genotechnology has thus far escaped any recognised major and spectacular disasters.

## What future Aladdin?

### *The technological 'promise' of environmental hazards*

The genotech future, we are assured, is in the global society's best interest: genetic engineering is the only means to provide unlimited amounts of cheap food for ever-rising numbers of people. With nuclear technology the same argument was made about energy. Today the energy promise shows little hope of being fulfilled whereas the neglected hazard potential and the cost of decommissioning of this technology have risen beyond the innovators' worst nightmare scenarios. Genetic engineering, we are told, is the only means to satisfy not just today's but future socio-environmental needs in a safe and environmentally benign way. This too was promised by the promoters of nuclear power and the chemical management of the environment. It too turned out to be an unfulfilled and unfulfillable dream. In addition to these similarities between genotechnology and predecessor technologies, however, there are a number of pertinent differences that require our attention.

Whatever accidental changes may have been brought about so far by genetically engineered products, they lack the drama, immediacy and certainty of a Bhopal or Chernobyl accident. This absence of attributable and/or publicly acknowledged disasters tends to get confused with safety. Yet, as I have shown in this analysis, it does not mean that the hazards do not exist. It does not signify safety and should under no circumstances lull citizens into a false sense of security. Like nuclear and chemical materials, genetically modified 'products' act invisibly at the level of cells. But their effects and symptoms, other than those intended by the specific genetic manipulation, are likely to be subtle rather than dramatic and/or not show themselves for many years. Moreover, the genetically altered materials' freedom vastly extends that of predecessor innovations. As living processes they multiply, mutate, migrate, and spread invisibly and unnoticeably by means that fall outside the remit of their creation. Organisms can turn malignant and develop into a pandemic without this being immediately obvious, noticeable or detectable. Quite clearly, this dramatically increases the materials' hazard potential.

With genetically modified organisms (GMO's), for example, harmful effects are not just more difficult to predict but even more difficult to trace to their origins than is the case with radiation, hormone disrupting chemicals, or diseases such as BSE and CJD (see also Chapters 1, 4 and 6). With thousands of genetically engineered viruses, bacteria, plants and animals developed over the last twenty years, it is now a case of potential hazards in the wings, of the global community of beings – present and future – having to wait and see what physical and socio-environmental 'effects' will ensue over the next years, decades, centuries, millennia. When symptoms do arise there will be inescapable uncertainty about their causes. Proof will be unobtainable, a thing of the past, a luxury of a bygone era. The only certainty in this context of indeterminacy is the fact that released GMOs cannot be recalled, that they cannot even be traced. They have become free agents, instantaneous new life forms, ready to interact with others whose evolutionary pedigree extends over millennia.

Opponents of genotechnology tend to view genetically modified 'products' as scientific monstrosities, as modern Frankensteins. Proponents, on the other hand, prefer to liken GMOs to the colonial period when plants and animals were brought back from foreign places and introduced into a new context. While this Kew Gardens image of GMOs is intended to allay any fears and reservations people might have, it may equally conjure up memories of mutant killer bees and numerous other introductions of species from foreign lands that turned into disasters (for examples of this, see Rosenbladt, 1988: Chapter 5). The lessons to be learnt from such introductions of alien species are unambiguously clear: species co-evolve with other species and their environments as ecosystems. There are no cut-off points for co-evolutions. *The temporal web of interdependencies reaches to the beginning and the end of time.* If you take something out of such a system, then all its other dependencies are implicated. If you add something, there is no way of predicting how that which has not co-evolved into a network of inter-dependencies will fare as an abstract, decontextualised, instantaneous entity that has to establish for itself a context for future networked relations of co-evolution. The organisms' past behaviour and response patterns may or may not be appropriate for its present and future survival. It may thrive, take over, go under or barely manage to exist.

Introduction of a new organism into the environment is precisely what genotechnology is about. Through manipulation at the level of genes this scientific skill and technique is used to develop novel micro-organisms for remedial purposes: for breaking down oil slicks and digesting crude oil; for combating rabies in foxes and other such designated threats to human health; for dealing with toxic chemicals in the environment and on waste dumps, even explosive substances and radioactive waste. While most of the hazards and their 'treatments' are real enough, there is no way of telling whether or not the 'remedy' will in due course become yet another item on the ever increasing list of hazards.

## *The social hazard of genetically engineered futures*

Hazards of a different, that is, a socio-political kind are created with the patenting of genetically modified organisms that range from micro-organisms to plants and animals, including their parts and processes. The current drive towards patenting emanates from the US, with European and other industrialised countries not trailing too far behind. According to the European Patent Office, only genetically fixed plants are excluded from potential patenting. This means that it is just a question of time before all plants and animals are 'protected' under some patent or other, that is, before the entirety of natural reproduction has come under the control of a few transnational corporations (see also Leskien, 1994:45). To Vandana Shiva (1993:122), 'this is just another way of stating that global monopoly over agriculture and food systems should be handed over as a right to multi-national corporations'. It is the commodification and privatisation of life itself. As such it is not merely redrawing the lines of inequality but reviving some exceedingly unsavoury socio-political and medical practices, especially where these have the potential to be combined. An example would be the conjoining of patenting and cloning.

At present, humans are excluded from patenting and cloning. Yet, the processes if not the patents and policies are in place to screen for defects and disability and to clone humans with the most desirable and coveted traits. The scientific and medical skills are in place to ensure a future of perfect people, a future where all those who might turn out to be a drain on collective resources can be excluded either from insurance cover or, more sensible still, from existence. We seem to have been here before. In a previous political system, people with Huntington's disease, mental disabilities and a host of other 'abnormalities' ended up in the gas chamber or at the very least had enforced sterilisations. Freely translated from Sabine Rosenbladt (1988:242), the idea was that 'the collective had to be freed from bad hereditary traits, that the cost was too high for society to support an army of diseased, feeble-minded, cripples, deaf, blind and criminals'. For some forty years there was a moratorium on such ideas but now, it seems, the distance of one generation is sufficiently large not to associate the current brave new world of genotechnology-based 'preventative' medicine with the political system that developed such social eugenics to perfection.

With genetically engineered foods and medicines too, the practices and patterns of exploitation have a familiar ring. As with previous food experiments and introductions of new medical technologies (see Chapter 5), the powerless have been chosen to act as guinea pigs for the latest genetically engineered inventions: to test the 'cures', grow the plants, rear the animals, drink their milk, eat the food. With their substantially lower life-expectancy, the people testing these products could conveniently succumb to hunger, disease or strife long before the effects of those experimental foods and medicines show up as symptoms. This means, with a bit of luck – for the companies 'testing' their products – people and laboratory animals will be dead before the genetically modified nourishments and cures emerge from the latent stage as symptom, that is, before they are recognised as pathogens. What, we need to ask, can testing mean in such a time–space distantiated context? How many generations, of what beings, over what spatial range, have to be clear of adverse symptoms for a product to be safe? How are latent and immanent multiplications, mutations, syntheses and migrations to be established and measured? How are pathogens to be traced to their origins?

How is preclinical harm to be identified? And so the socio-scientific, socio-legal and socio-political questions go on.

### The stakes are high, regulations meagre, public information non-existent

Where genetically modified foods are concerned, the stakes are high in a dual way. First, genotechnology holds out the promise to grow artificially in the North what currently is provided by the South. Second, the scale of financial investments by transnational corporations and the stock market in genetic research and development is unprecedented. Disguised as food for the rising populations of the future, the business giants of the industrialised North are currently busying themselves with the creation of crops that will substitute cash crops grown at present by the South. They are developing plants that can survive harsher climatic conditions: designer plants that get genetically transformed into substitutes for existing crops such as alternatives to sugar; rape-seed oil that gets changed into olive, palm or coconut oil; palm oil that, in turn, gets redesigned into cocoa-butter. There seems in principle no limit to the possible mutations (see Spangenberg, 1994; Rosenbladt, 1988). But, as Vandana Shiva points out,

> The impact of successful production of substitutes will be felt most by countries which have, in an earlier international division of labour, been made dependent upon export of the natural products concerned. This will be particularly destructive to the economies of Africa which depend entirely on single crops for most of their export earnings. While historically Africa was used to growing crops needed for Europe, in the emerging world order based on new biotechnologies, Africa will become dispensable as the North finds biotech substitutes for African crops.
>
> (Shiva, 1993:117)

This development should not be taken as a sudden conversion by transnational corporations: they are not redeveloping national loyalties (see Chapters 2, 3 and 4). It is simply yet another innovative way to create new business opportunities and to pursue the competitive edge and financial fitness.

Genotechnology, I want to suggest, is far more difficult to resist and fight than was/is nuclear power. Almost irrespective of whether or not there is wide-ranging public recognition of its dangers, genotechnology is being pursued to some indefinite bitter future (rather than end). The reasons behind this are the size of the financial investment and the fact that, unlike the development of the nuclear sector, most of this investment and production is private or corporate. Although sensitive to consumer pressure, corporations are neither dependent on the vote for their survival nor bound to national boundaries for their activities and outlets. Moreover, as decades of development come to 'fruition', any kind of retraction is out of the question, off the agenda, not open for negotiation. The result has to work at all cost as experience with earlier so-called wonder drugs such as human insulin, clotting factor and growth hormone has shown (for details see Krimsky, 1991: 26–31; Rosenbladt, 1988:189–92).

> The failure of biotechnology was out of the question whether or not there were existing needs or favourable markets. The techniques would be used over and over again until successful products emerged. If the first generation of products proved unsuccessful or only marginally successful, other needs would be found to support a second generation of products.
>
> (Krimsky, 1991:31)

The cancers and deaths that arose from some of these treatments were temporally close and direct enough for scientists and doctors to establish causal links. No such connections, however, are likely for the long-term changes at the molecular and genetic level as they work their way through bodies and systems. As I argued in Chapter 1, the problems that are recognised tend to be those of a previous technology. It seems a mark of true innovation that the real dangers associated with it still fall outside the involved scientists' imaginative capacity. Moreover, the extremely mechanistic assumptions underpinning genotechnology predispose its activists against being able and/or inclined to conceive of the potential effects of this technology's application in post-Newtonian terms. Thus, I agree with Regine Kollek (1995a:105) when she argues that 'the methods are both megaphone and hearing aid of scientific research'.

This clearly is a difficult context for policy and regulation. Governments operating within frameworks of four to five year election cycles, concerned to boost their country's GNP and export, and keen to see jobs for the future are naturally drawn to the lure of this technology. It promises to fulfil all those political needs and more. And it is safe – within the time-scale of politics, that is. Moreover, most major forms of regulation relating to health hazards and environmental issues arise from enforced responses to disasters or crises. The high level of invisibility and the absence of dramatic accidents or large-scale disasters, therefore, make genotechnology a poor candidate for regulation.

Initially, unfamiliarity brought about cautious restrictions. With the dramatic increase in genotechnological activity, its massive contribution to the national economy, and its assumed 'safety record', these early concerns are being gradually but surely whittled away. However, some regulating authorities are more inclined than others to apply the precautionary principle. The EU is an interesting case in point. One of the least democratically constituted regulatory structures, the EU, is actually more cautious than most other Liberal Democracies (see also Chapter 3). The continued moratorium on growth hormones in milk is a case in point. Here the precautionary principle is applied for fear of collapse of the dairy and meat market. This throws up an interesting paradox: whilst concerned citizens and environmental pressure groups are still conceived as agitators and troublemakers, the threat of a consumer boycott seems to work wonders. True to economic form and the ideology that underpins the globally constituted Liberal Democracies, who lead the way with their commitment to genotechnological developments, consumers are taken seriously by their political leaders. Globally constituted transnational corporations, in contrast, can ignore nationally located consumers as long as they are assured of the continued commitment by politicians, scientists, and the medical profession to the genotechnology cause. Together, of course, governments and transnational corporations have nothing but society's best interest at heart.

Finally, in the absence of major disasters, genotechnology has not had the breadth of press coverage and the depth of media analysis that would be necessary to facilitate the beginning of the kind of public genetic literacy I mentioned earlier. Information, analysis and debate at the popular level is virtually non-existent. Neither media hyperactivity around the cloning of sheep nor academic debates late at night qualify for the sort of public dissemination of information and debate that is necessary for citizens to grasp the changes that are currently being set in place in the name of humanity's future.

At the moment, only the people who are directly affected are forced into an almost instantaneous 'coming up to speed' with the technology, its potential, limits and dangers. Offered genetic counselling or diagnosis of self, relatives or offspring, for example, recipients of this attention cannot avoid engagement with issues that will affect their long-term future. Faced with a decision of whether or not to get involved in hormone treatment of their children or livestock, people are forced to find out as much as they can before they come to a decision. It is extremely unlikely that they would get any of that information from the news media, especially from the press, since only exotic claims, monstrous results, or disastrous effects will claim journalists' focused attention. Only in that form is the mediagenecy of genotechnology assured. Television and radio, in the UK at least, have offered some investigative programmes on food and medical issues surrounding the new technology but these were few and far between, late at night and on elite channels – certainly not enough to create widespread social awareness. Moreover, science programmes such as 'Tomorrow's World' which tend to marvel uncritically at science's achievements are more likely than not to have strengthened expectations that genotechnologies are on the verge of providing the solutions to all current ills and troubles, and to all future needs of humans and 'their' environment.

On the basis of the above, it may look as if hazard-creation was an inevitable, foregone conclusion, but this is not the case. A change of direction may be extremely difficult to achieve, but it is certainly not impossible, and the temporal angle of vision gives us previously unnoted pointers to alternative paths. It certainly leaves us in no doubt that the problems and disasters associated with previous technological innovations cannot provide the necessary and sufficient knowledge about the kind of problems that will arise from these latest developments. At the same time, their previous unpredictability and their continuous production of 'surprises' leaves no doubt about the need for caution and precaution for socio-environmental situations where the predictability afforded by the laboratory abstraction is lost and/or forfeited at the very moment the creation is set free to interact, transact, migrate, and move at an indeterminate, variable pace through the system as a whole. Thus, not the knowledge of the past, not the certainties of a previous age, not the science of proof and certainty, but proper engagement with temporality, indeterminacy and time–space distantiation have to become key issues for praxis. In additon, leaps of the imagination are necessary for any meaningful projection into the future of systems that have outgrown the scientific framework of their creators.

In this book I have not merely put a different spin on 'the environmental problem' but through the timescape perspective I have sought to make the invisible visible, the im/material real. Focus on the time–space of environmental hazards helped to reformulate both 'problems' and 'cures' and guided attention to the processes below the surface, to the *Wirkwelt* of *natura naturans* that acts immanently for indefinite periods and over unspecified places. It enabled us to redefine the nature and interdependence of safety, risks and hazards, to recognise and articulate the need for new conceptual tools, metaphors, theories and information structures. It made real what is currently assumed non-existent. As such, the timescape perspective has the potential to facilitate the will to a sense and sensibility that befits the current hazard re/production of the industrial way of life.

*Plate 8* Cromlech, Dyffryn Ardudwy from *Wales the First Place* by Paul Wakefield. *Source*: Tony Stone Images.

# REFERENCES

Adam, B. (1988) 'Social versus Natural Time, a Traditional Distinction Re-examined', in Young, M. and Schuller, T. (eds) *The Rhythms of Society*, London: Routledge.

Adam, B. (1990) *Time and Social Theory*, Cambridge: Polity Press; Philadelphia: Temple UP.

Adam, B. (1992) 'Modern Times: The Technology Connection and its Implications for Social Theory', *Time & Society* 1,2:175–92.

Adam, B. (1993) 'Within and Beyond the Time Economy of Employment Relations: Conceptual Issues Pertaining to Research on Time and Work', *Social Science Information sur les Sciences Sociales* 32,2:163–84.

Adam, B. (1994/5) 'Time for Feminist Approaches to Technologies, "Nature" and Work', *Arena* 4:91–104.

Adam, B. (1995a), *Timewatch: The Social Analysis of Time*, Cambridge: Polity Press; Williston, VT: Blackwell.

Adam, B. (1995b) 'Auf dem Weg zur Laborzeit: Wandel der Zeiten in der Landwirtschaft', *Politische Ökologie*, Sonderheft 8:20–25.

Adam, B. (1995c) 'Von Urzeiten und Uhrenzeit: Eine Symphonie der Rhythmen des täglichen Lebens', in Held, M. and Geißler, K. (eds) *Von Rhythmen und Eigenzeiten: Perspektiven einer Ökologie der Zeit*, Stuttgart: Hirzel, Edition Universitas, pp. 19–30.

Adam, B. (1996a) 'Re-vision: The Centrality of Time for an Ecological Social Science Perspective', in Szerszynski, B., Lash, S. and Wynne, B. (eds) *Risk, Environment and Modernity: Towards a New Ecology*, London: Sage, pp. 84–103.

Adam, B. (1996b) 'The Technology–Ecology Connection and its Conceptual Representation', in Fraser, J. T. and Soulsby, M. P. (eds) *The Dimensions of Time and Life*, Madison, CT: International UP, pp. 207–14.

Adam, B. (1996c) 'Beyond the Present: Nature, Technology and the Democratic Ideal', *Time & Society* 5,3:319–38.

Adam, B. (1997a) 'Chernobyl: Implicate Order of Socio-environmental Chaos' in Fraser, J. T. and Soulsby, M. (eds) *Time and Chaos*, Madison, CT: International UP.

Adam, B. (1997b) 'Radiated Identities: In Pursuit of the Temporal Complexity of Conceptual Cultural Practices', in Featherstone, M. and Lash, S. (eds) *Spaces of Culture: City, Nation, World*, London: Sage, in press.

Adam, B., Held, M. and Geißler, K. eds (1997a) *Die Nonstop- Gesellschaft und ihr Preis*, Stuttgart: Universitas.

Adam, B., Held, M., Geißler, K., Kümmerer, K. and Schneider, M. (1997b) 'Time for the Environment: The Tutzing Project "Time Ecology"', *Time & Society*, 6,1:73–84.

Adam, B. and Kütting, G. (1995) 'Time to Reconceptualize 'Green Technology' in the Context of Globalization and International Relations', *Innovation* 8 (3):243–259.

Allan, S. (1997) 'News from NowHere: Televisual News Discourse and the Construction of Hegemony' in Bell, A. and Garrett, P. (eds) *Approaches to Media Discourse*, Oxford: Blackwell, pp. 105–41.

Altner, G. (1993) 'Der Natur ihre Zeit geben, ihre Zeit lassen', in Held, M. and Geißler, K. (eds) *Ökologie der Zeit. Vom Finden der rechten Zeit maße*, Stuttgart: Universitas, pp. 133–42.

Arendt, H. (1958) *The Human Condition*, Chicago: Chicago UP.

Arwood, M. (1993) 'Blind Faith and the Free Trade', in Nader, R. (ed.) *The Case against Free Trade*, San Francisco: Earth Island Press, pp. 92–6.

Aubrey, C. (1991) *Meltdown: The Collapse of the Nuclear Dream*. London: Collins and Brown extracted in Aubrey, C., Grunberg, D. and Hildyard, N. (eds) (1991) *Nuclear Power: Shut it Down! Vols. 1 and 2*, London: The Ecologist, pp. 136–9.

Aubrey, C., Grunberg, D. and Hildyard, N. (eds) (1991) *Nuclear Power: Shut it Down! Vols. 1 and 2*, London: The Ecologist.

Baker, S. (1996) 'Environmental Policy in the European Union: Institutional Dilemmas and Democratic Practice', in Laffarty B. and Meadowcroft J. (eds) *Environment and Democracy*, London: Edward Elgar, pp. 213–31.

Bartussek, H. (1995) 'Zeit der Tiere – Raum für Tiere. Die Haltung von Tieren in der Landwirtschaft', *Politische Ökologie*, Special Issue 8:66–70.

Beck, U. (1992a) *Risk Society: Towards a New Modernity*, London: Sage.

Beck, U. (1992b) 'From Industrial Society to Risk Society: Questions of Survival, Social Structure and Ecological Enlightenment', *Theory, Culture and Society* 9:97–123.

Beck, U. (1995a) *Ecological Enlightenment: Essays on the Politics of the Risk Society*, translated by Mark Ritter, New Jersey: Humanities Press.

Beck, U. (1995b) *Die feindlose Demokratie. Ausgewählte Aufsätze*, Stuttgart: Reclam, pp. 182–92.

Beck, U. (1996) 'Risk Society and the Provident State', in Szerszynski, B., Lash, S. and Wynne, B. (eds) *Risk, Environment and Modernity: Towards a New Ecology*, London: Sage, pp. 27–43.

Beck, U., Giddens, A. and Lash, S. (1994) *Reflexive Modernisation. Politics, Tradition and Aesthetics in the Modern Social Order*, Cambridge: Polity.

Bell, A. (1995) 'News Time', *Time & Society* 4,3:305–328.

Benton, T. and Redclift, M. (1994) 'Introduction' to Redclift, M. and Benton, T. (eds) *Social Theory and the Global Environment*, London: Routledge, pp. 1–28.

Bertell, R. (1985) *No Immediate Danger: Prognosis for a Radioactive Earth*, London: Women's Press.

Biervert, B. and Held, M. (eds) (1994) *Das Naturverständnis der Ökonomik. Beiträge zur Ethikdebatte in den Wirtschaftswissenschaften*, Frankfurt, a. M.: Campus.

Biervert, B. and Held, M. (eds) (1995) *Zeit in der Ökonomik. Perspektiven für die Theoriebildung*, Frankfurt, a. M.: Campus.

Biervert, B. and Held, M. (eds) (1996) *Die Dynamik des Geldes. Über den Zusammenhang von Geld, Wachstum und Natur*, Frankfurt, a. M.: Campus.

Biesecker, A. (1994) 'Wir sind nicht zur Konkurrenz verdammt. Auf der Suche nach alten und neuen Formen kooperativen Wirtschaftens' *Politische Ökologie*, Special Issue 6: 28–31.

Binswanger, H. C. (1991) *Geld und Natur – Das wirtschaftliche Wachstum zwischen Ökonomie und Ökologie*, Stuttgart: K. Thienemanns Verlag.

Binswanger, H. C. (1994) 'Geld und Natur', in Biervert, B. and Held, M. (eds) *Das Naturverständnis der Ökonomik. Beiträge zur Ethikdebatte in den Wirtschaftswissenschaften*, Frankfurt, a. M.: Campus, pp. 175–88.

Blythman, J. (1996) *The Food we Eat,* London: Michael Joseph.

Boden, D. (1997) 'Worlds in Action: Information, Instantaneity and Global Futures Trading', in Adam, B., Beck, U., v. Loon, J. and Taraborelli, P. (eds) *Positioning Risk,* London: Sage, in press.

Bohm, D. (1983) *Wholeness and the Implicate Order,* London: ARK.

Böhmer-Christiansen, S. and Skea, J. (1991) *Acid Politics: Environmental and Energy Policies in Britain and Germany,* London: Belhaven Press.

Bongaerts, J. C. (1994) 'Weiterhin Verhandlungsmasse. Konsequenzen für die Umweltpolitik der EU nach dem Vertrag von Maastricht', *Politische Ökologie,* 'Europa hat die Wahl', 37:34–7.

Briggs, J. P. and Peat, F. D. (1985) *Looking Glass Universe: The Emerging Science of Wholeness,* London: Fontana.

Brown, E.G. Jr. (1993) 'Free Trade is not Free', in Nader, R. (ed.) *The Case against Free Trade,* San Francisco: Earth Island Press, pp. 65–9.

Bunyard, P. (1988) 'Nuclear Energy after Chernobyl', in E. Goldsmith and N. Hildyard (eds) *The Earth Report: Monitoring the Battle for our Environment,* London: Beazley.

Busch-Lüty, C. (1994) 'Ökonomie als Lebenswissenschaft. Der Paradigmenwechsel zum Nachhaltigkeitsprinzip als wissenschaftstheoretische Herausforderung', *Politische Ökologie,* Special Issue 6:12–17.

Busch-Lüty, C. and Dürr, H.P. (1993) 'Ökonomie und Natur: Versuch einer Annäherung im Interdisziplinären Dialog', in König, H. (ed.) *Umweltverträgliches Wirtschaften als Problem von Wissenschaft und Politik.* Berlin: Duncker & Humblot, pp. 13–44.

Byrne, C. (1996) 'Genetic Boost makes our Cox's Crisper', *Independent,* 1/9/1996:8.

Cairncross, F. (1995) *Green Inc.: A Guide to Business and the Environment,* London: Earthscan.

Colborn, T., Meyers, J. P. and Dumanoski, D. (1996) *Our Stolen Future: How Man-made Chemicals are Threatening our Fertility, Intelligence and Survival,* Boston: Little, Brown & Company.

Davies, M. (1996) 'Cosmic Dancers on History's Stage? The Permanent Revolution in the Earth Sciences', *New Left Review* 217:49–84.

Derrida, J. (1984) 'No Apocalypse, Not Now (Full Speed Ahead, Seven Missiles, Seven Missives)' *Diacritics* 14:20–31.

Dörner, D. (1989) *Die Logik des Misslingens,* Hamburg: Rowohlt.

Elliott, T. (1990) 'Food Irradiation – Miracle or Menace?' *Conservation Now* 1,1:42–5.

Energy Economist (1990) 'World Status; Chernobyl Revisited', *Energy Economist* 103 extracted in Aubrey, C., Grunberg, D. and Hildyard, N. (eds) 1991 *Nuclear Power: Shut it Down! Vols. 1 and 2,* London: The Ecologist, pp. 144–9.

Ermarth, E. D. (1992) *Sequel to History: Postmodernism and the Crisis of Representational Time,* Princeton, N.J.: Princeton UP.

Esteva, G. and Prakash, M. S. (1994) 'From Global to Local Thinking' *The Ecologist* 24:162–4.

Florenz, K. (1994) 'Das Recht Ja oder Nein zu sagen. Das Europäische Parliament nach Maastricht und seine Bedeutung für den Umweltschutz', *Politische Ökologie,* 'Europa hat die Wahl', 37:11–13.

Gebers, B. (1994) 'Kaum zukunftsweisende Ansätze. Bewertung der Parteiprogramme aus der Sicht des Umweltrechts', *Politische Ökologie,* 'Europa hat die Wahl', 37:32–3.

George, S. (1984) *Ill Fares the Land: Essays on Food Hunger and Power,* London: Writers and Readers Publishing Cooperative Society Ltd.

Gibbons, M., Limoges, C., Nowotny, H., Schwartzman, S., Scott, P. and Trow, M. (1994) *The New Production of Knowledge: The Dynamics of Science and Research in Contemporary Societies,* London: Sage.

233

Giddens, A. (1991) *Modernity and Self-Identity: Self and Society in the Late Modern Age*, Cambridge: Polity.

Giddens, A. (1994a) *Beyond Left and Right: The Future of Radical Politics*, Cambridge: Polity.

Giddens, A. (1994b) 'Living in a Post-Traditional Society', in Beck, U., Giddens, A. and Lash, S. *Reflexive Modernisation: Politics, Tradition and Aesthetics in the Modern Social Order*, Cambridge: Polity, pp. 56–109.

Gleick, J. (1987) *Chaos: Making a New Science*, London: Penguin.

Goodman, D. and Redclift, M. R. (1991) *Bibliotextashioning Nature: Food , Ecology and Culture*, London: Routledge.

Gourlay, K. A. (1992) *World of Waste: Dilemmas of Industrial Development*, London: Zed Books.

Grove-White, R. (1996) 'Environmental Knowledge and Public Policy Needs: On Humanising the Research Agenda' in Szerszynski, B., Lash, S. and Wynne, B. (eds) *Risk, Environment and Modernity: Towards a New Ecology*. London: Sage, pp. 269–86.

Harding, G. (1968/1992) 'The Tragedy of the Commons' reprinted in Markandya, A. and Richardson, J. (eds) *The Earthscan Reader in Environmental Economics*, London: Earthscan.

Hayles, K. (1990) *Chaos Bound: Orderly Disorder in Contemporary Literature and Science*, Ithaka: Cornell UP.

Heidegger, M. (1927/1980) *Being and Time*, translated from the German by Macquarrie, J. and Robinson, E., Oxford: Blackwell.

Heiland, S. (1992a) *Naturverständnis. Dimensionen des menschlichen Naturbezugs*, Darmstadt: Wissenschaftliche Buchgesellschaft.

Heiland, S. (1992b) 'Mißverstandene Natur. Aspekte unseres Alltagsverständnisses von Natur', *Politische Ökologie*, 'Entfremdete Natur', 25:46–52.

Held, M. and Geißler, K. (eds) (1993) *Ökologie der Zeit. Vom Finden der rechten Zeitmaße*, Stuttgart: Universitas.

Held, M. und Geißler, K. (eds) (1995) *Von Rhythmen und Eigenzeiten. Perspektiven einer Ökologie der Zeit*, Stuttgart: Universitas.

Hewett, J. (ed.) (1995) *European Environmental Almanac*, London: Earthscan.

Hey, C. (1994) 'Das Spiel von Hase und Igel. Vorreiter und Bremser der europäischen Umweltpolitik', *Politische Ökologie*, 'Europa hat die Wahl', 37:7–10.

Hofmeister, S. (1994) 'Auf dem Weg in eine nachhaltige Stoffwirtschaft. Über die Chancen einer Wiederentdeckung der physischen Reproduktion durch die industrielle Witschaftsgemeinschaft', *Politische Ökologie*, Special Issue 6:51–5.

Hofmeister, S. (1997) 'Nature's Temporalities: Consequences for Environmental Politics', *Time & Society*, 6,2: 309–22.

Ingold, T. (1993) 'The Temporality of Landscape', *World Archaeology*, 25:152–74.

Inhetveen, H. (1994) 'Hortikultur als Vorbild. Am Beispiel der Nutzgartenwirtschaft können wichtige Aspekte des vorsorgenden Wirtschaftens entfaltet werden' *Politische Ökologie*, Special Issue 6:22–7.

Jacobs, M. (1991) *The Green Economy: Environment, Sustainable Development and the Politics of the Future*, London: Pluto Press.

Jacobs, M. (1996) *The Politics of the Real World: Meeting the New Century*, London: Earthscan.

Kern, S. (1983) *The Culture of Time and Space 1880–1918*, London: Weidenfeld and Nicolson.

Kluge, T. and Schramm, E. (1995) 'In die Tiefe der Zeit. Nachhaltiger Umgang mit dem Lebensmittel Wasser', *Politische Ökologie*, Special Issue 8:39–43.

Kollek, R. (1995a) 'The Limits of Experimental Knowledge: A Feminist Perspective on the Ecological Risks of Genetic Engineering', in Shiva, V. and Moser, I. (eds) *Biopolitics: A Feminist and Ecological Reader on Biotechnology*, London/New Jersey: Zed Books, pp. 95–111.

Kollek, R. (1995b) 'Zeit der Natur – Zeit der Kultur. Zum Verhältnis von Evolution, Züchtung und Gentechnik', *Politische Ökologie*, Special Issue 8:25–30.

Korten D. C. (1995) *When Corporations Rule the World*, London: Earthscan.

Krimsky, S. (1991) *Biotechnics and Society: The Rise of Industrial Genetics*, New York: Praeger.

Lacey, R. W. (1994) *Mad Cow Disease: The History of BSE in Britain*, Jersey, Channel Islands: Gypsela.

Latour, B. (1993) *We have never been Modern*, translated by C. Porter, London: Harvester.

Leckie, S. (1995) 'Housing as Social Control in Tibet', *The Ecologist* 25,1:8–11.

Leskien, D. (1994) 'Patent auf das Leben. Der Einbruch des Patentrechts in die Biologie', *Politische Ökologie*, 'Geniale Zeiten' 35:43–6.

Lovelock, J. (1979) *Gaia: A New Look on Life on Earth*, Oxford: Oxford UP.

Lovelock, J. (1988) *The Ages of Gaia: A Biography of our Living Earth*, Oxford: Oxford UP.

Luce, G. G. (1977) *Body Time: The Natural Rhythms of the Body*, St Albans: Paladin.

Luhmann, N. (1978) 'Temporalisation of Complexity', in Geyer, R. F. and Zouwen, J. v.d. (eds) *Socio-cybernetics: An Actor-oriented Social Systems Approach*, London: Martinus Nijhoff, pp. 92–113.

Luhmann, N. (1982a) 'The Future cannot Begin: Temporal Structures in Modern Society', in *The Differentiation of Society*, translated from the German by S. Holmes and C. Larmore, New York: Columbia UP, pp. 271–89.

Luhmann, N. (1982b) 'World-time and System History', in *The Differentiation of Society*, translated from the German by S. Holmes and C. Larmore, New York: Columbia UP pp. 289–324.

Mander, J. (1993) 'Megatechnology, Trade and the New World Order', in Nader, R. (ed.) *The Case against Free Trade*, San Francisco: Earth Island Press, 13–22.

Marx, K. (1932/1968) *The German Ideology*, Moscow: Progress Publishers.

Marx, K. (1857/1973) *Grundrisse*, Harmondsworth: Penguin.

May, J. (1989) *The Greenpeace Book of the Nuclear Age: The Hidden History; the Human Cost*, London: Gollancz, also extracted in Aubrey, C., Grunberg, D. and Hildyard, N. (eds) (1991) *Nuclear Power: Shut it Down! Vols. 1 and 2*, London: The Ecologist, pp. 129–34.

McNair, B. (1988) *Images of the Enemy*, London: Routledge.

Mead, G. H. (1939/1952) *The Philosophy of the Present*, LaSalle, Ill.: Open Court.

Medvedev, Z. (1990) *The Legacy of Chernobyl*, Oxford: Blackwell.

Melbin, M. (1987) *Night as Frontier: Colonizing the World after Dark*, New York: Free Press, Macmillan.

Misch, A. (1994) 'Assessing Environmental Health Risks', in Brown, L. *et al.* (eds) *State of the World 1994*, New York: W.W. Norton & Co, pp. 117–37.

Nader, R. (1993) 'Free Trade and the Decline of Democracy' in Nader, R. (ed.) *The Case against Free Trade*, San Francisco: Earth Island Press, pp. 1–12.

*The New Internationalist* (1995, April) 'Nomads at the Crossroads', No. 266.

Nowotny, H. (1989/1994) *Time: The Modern and Postmodern Experience*, translated from the German by Neville Plaice, Cambridge: Polity.

*Politische Ökologie* (1994) Busch-Lüty, C., Jochimsen, M., Knobloch, U. and Seidl, I. (eds) Special Issue 6: 'Vorsorgendes Wirtschaften. Frauen auf dem Weg zu einer Ökonomie der Nachhaltigkeit'.

*Politische Ökologie* (1995) Schneider, M., Geißler, K. and Held, M. (eds) Special Issue 8: 'Zeit-Fraß. Zur Ökologie der Zeit in Landwirtschaft und Ernährung'.

Postler, G. (1995) 'Lebens- oder Höchstleistung? Vom Hirten zum Gentechniker in der Tierzucht', *Politische Ökologie*, Special Issue 8:57–60.

Price, C. (1993) *Time, Discounting and Value*, Oxford: Blackwell.

Prigogine, I. and Stengers, I. (1984) *Order out of Chaos: Man's New Dialogue with Nature*, London: Heinemann.

Quarrie, J. (ed.) (1992) *Earth Summit 1992, The United Nations Conference on Environment and Development*, London: Regency Press.

Rifkin, J. (1985) *Declaration of a Heretic*, Boston/London: Routledge and Kegan Paul.

Rifkin, J. (1987) *Timewars: The Primary Conflict in Human History*, New York: Henry Holt.

Rifkin, J. (1992) *Beyond Beef: The Rise and Fall of the Cattle Culture*, London: Thorsons.

Rifkin, J. and Howard, T. (1985) *Entropy: A New World View*, London: Paladin.

Roche, P. (1990) 'The Legacy of Chernobyl', *SCRAM* 77:11, also extracted in Aubrey, C., Grunberg, D. and Hildyard, N. (eds) (1991) *Nuclear Power: Shut it Down! Vols.1 and 2*, London: The Ecologist, pp. 150–1.

Romanyshyn, R. D. (1989) *Technology as Symptom and Dream*, London: Routledge.

Rose, K. J. (1989) *The Body in Time*, New York: John Wiley & Sons.

Rosenbladt, S. (1988) *Biotopia. Die genetische Revolution and ihre Folgen für Mensch, Tier und Umwelt*, München: Knaur.

Ross, A. (1994) *The Chicago Gangster Theory of Life: Nature's Debt to Society*, London: Verso.

Scheinost, A. (1995) 'Entstehen, wachsen und vergehen. Von den Zeitmaßen der Resource Boden' *Politische Ökologie*, Special Issue 8:35–8.

Schmidheiny, S. with the Business Council for Sustainable Development (1992) *Changing Course: A Global Business Perspective on Development and the Environment*, Cambridge (US): MIT Press.

Schneider, M., Geissler, K. and Held, M. (eds) (1995) 'Zeit-fraß. Zur Ökologie der Zeit in Landwirtschaft und Ernährung' *Politische Ökologie*, Special Issue 8:6–14.

Schneider, M. (1997) 'Tempo Diet', *Time & Society*, 6,1:85–98.

Shama, S. (1995) *Landscape and Memory*, London: Harper Collins.

Sheldrake, R. (1983) *A New Science of Life*, London: Paladin.

Sheldrake, R. (1990) *The Rebirth of Nature: The Greening of Science and God*, London: Century.

Shiva, V. (1993) *Monocultures of the Mind: Perspectives on Biodiversity and Biotechnology*, London: Zed Books.

Shiva, V. and Moser, I. (eds) (1995) *Biopolitics: A Feminist and Ecological Reader on Biotechnology*, London/New Jersey: Zed Books.

Simmel, G. (1907/1978) *The Philosophy of Money*, London: Routledge and Kegan Paul.

Soper, K. (1995) *What is Nature? Culture, Politics and the Non-Human*, Oxford: Blackwell.

Spangenberg, J. (1994) 'Die Speisung der 4000 Millionen. Gentechnik: Wunderwaffe gegen Welthunger?' *Politische Ökologie*, 'Geniale Zeiten' 35:58–62.

Szerszynski, B., Lash, S. and Wynne, B. (eds) (1996) *Risk, Environment and Modernity. Towards a New Ecology*, London: Sage.

Tansey, G. and Worsley, T. (1995) *The Food System: A Guide*, London: Earthscan.

Taschner, K. (1994) 'Ansatzweise richtig. Die Umweltpolitik der europäischen Gemeinschaft' *Politische Ökologie*, 'Europa hat die Wahl', 37:2–4.

Telkämper, W. (1994) 'Zentral oder dezentral? Wie europäische Umweltpolitik erfolgsversprechend gestaltet werden kann', Special Issue, 'Europa hat die Wahl', 37:14–15

Trommer, G. (1988) *Ökologisch bedeutsame Naturvorstellungen in deutschen Bildungskonzepten. Habilitationsschrift im Fachbereich Landespflege der Universität Hannover*, unpublished.

Uexküll, J. v. and Kriszat, G. (1934) *Streifzüge durch die Umwelten von Tieren und Menschen*, Frankfurt, a. M.: Fischer.

Wallach, L. (1993) 'Hidden Dangers of GATT and NAFTA', in Nader, R. (ed.) *The Case against Free Trade*, San Francisco: Earth Island Press, pp. 23–64.

Weber, M. (1904–5/1989) *The Protestant Ethic and the Spirit of Capitalism*, London: Unwin and Hyman.

Weber, M. (1919/1985) 'Science as a Vocation', in Gerth, H. H. and Mills, C. Wright (eds) *From Max Weber: Essays in Sociology*, London: Routledge and Kegan Paul, pp. 129–58.

Weber, M. (1949/1985) *From Max Weber: Essays in Sociology*, Gerth, H. H. and Mills, C. Wright (eds) London: Routledge and Kegan Paul.

Weizsäcker, E. U. v. (1994) *Earth Politics*, London: Zed Books.

Welsh, I. (1994) 'Letting the Research Tail Wag the End-user's Dog: The Powell Committee and UK Nuclear Technology, *Science and Public Policy*, 21,1:43–53.

Wheale, P. and McNally, R. (1988) *Genetic Engineering: Catastrophe or Utopia?* Hemel Hempstead: Harvester Wheatsheaf.

Wheale, P. and McNally, R. (1995) *Anima Genetic Engineering: Of Pigs, Oncomice and Men*, London: Pluto Press.

White Jr., L. (1962) *Medieval Technology and Social Change*, Oxford: Oxford UP.

Williams, R. (1976) *Key Words: A Vocabulary of Culture and Society*, Glasgow: Fontana.

World Commission on Environment and Development (1987) *Our Common Future*, Oxford: Oxford UP.

Wynne, B. (1992) 'Misunderstood Misunderstanding: Social Identities and the Public Uptake of Science', *Public Understanding of Science* 1 (3):281–304.

Wynne, B. (1996) 'May the Sheep Safely Graze? A Reflexive View of the Expert–Lay Knowledge Divide', in Szerszynski, B., Lash, S. and Wynne, B. (eds) *Risk, Environment and Modernity: Towards a New Ecology*, London: Sage, pp. 44–83.

# INDEX